Celestial Mechanics

The Waltz of the Planets

Alessandra Celletti and Ettore Perozzi

Celestial Mechanics

The Waltz of the Planets

Published in association with
Praxis Publishing

Professor Alessandra Celletti
Università di Roma 'Tor Vergata'
Rome
Italy

Dr Ettore Perozzi
Telespazio
Rome
Italy

SPRINGER–PRAXIS BOOKS IN POPULAR ASTRONOMY
SUBJECT *ADVISORY EDITOR*: John Mason B.Sc., M.Sc., Ph.D.

ISBN 10: 0-387-30777-X Springer Berlin Heidelberg New York
ISBN 13: 978-0-387-30777-0 Springer Berlin Heidelberg New York

Springer is a part of Springer Science + Business Media (*springeronline.com*)

Library of Congress Control Number: 2006926301

Cover design: Jim Wilkie
Copy editing and graphics processing: R.A. Marriott
Typesetting: BookEns Ltd, Royston, Herts., UK

Printed in Germany on acid-free paper

Table of contents

To my husband Enrico and to my mother Mirella
Alessandra Celletti

To Eugenia Pennylane and Emilia Pollyjean
Ettore Perozzi

Foreword

I was delighted to be invited by my colleagues Alessandra Celletti and Ettore Perozzi to provide a foreword to their book, *Celestial Mechanics: The Waltz of the Planets*. Having known them for many years and long admired their work in the subject so many of us love and are fascinated by, I read with great attention and pleasure the text when it arrived. It is a formidable task they have set themselves, to provide a book that describes attempts by successive generations of astronomers from the dawn of history five millennia ago to observe, record and understand the phenomena of the heavens, particularly the intricate and perplexing behaviour of the planets, Sun and Moon. As naked eye astronomy became aided by the telescope and the photographic plate, and since the middle of the twentieth century, by instruments launched on spacecraft into circum-Earth orbit or to the Moon and planets and beyond, the discovery of new satellites, scores of them, and ring systems displaying new and initially perplexing behaviour also demanded explanations for that behaviour.

It is also the inspiring story of science itself with special reference to how lonely individuals, impelled by curiosity and dedicated to seeking the truth, and nothing but the truth, about the fascinating phenomena of nature, ultimately became accepted as scientists, those players in the most successful endeavour ever engaged in by the human race. It is the story of how their struggles ultimately prevailed against an entrenched and arrogant authority which believed it and it alone knew what to have faith in and would persecute and threaten (cf Galileo and his statement that the Earth orbited the Sun – did he really say under his breath: *eppur si muove* as he backed out?) and kill anyone displaying even more heretical tendencies (cf Giordano Bruno and his concept of the plurality of worlds). And it is the story how by the second half of the nineteenth century the immense prestige to the international brotherhood – and sisterhood – of scientists worldwide, won by their successes in pure science and technology, almost turned many of them into a new set of priests who were in danger of pronouncing that really before very long they would have discovered all of nature's secrets. Life would then hold nothing but carrying out experiments to add a few more figures to our measures of the fundamental constants of nature such as the constant of gravitation. How wrong can you be? Even then Einstein, Bohr, J.J.Thomson, and Rutherford, to mention only a few, were waiting in the wings preparing by their researches in sub-atomic physics to demolish the new establishment belief in the achievement of absolute truth.

This new establishment contained a harder type of scientist. To them, the long march of science had rescued much of the human race from tormented eras of superstition and fear that produced the horrific deaths of millions of people judged to be witches or religious heretics. The modern misuse of scientific and technological discoveries by our greedy, feckless society is quite another matter, a problem that must lead to disastrous worldwide consequences for humanity if twenty-first century society fails to solve it and is fast running out of time in which to do so.

The popular myth of a scientist is of a rational person who observes, notes, produces a theory or hypothesis, and carefully sets up an experiment to verify or to disprove that theory. If the result of the experiment supports the theory, the scientist has greater faith in his theory, especially if studies by other scientists, replicating his experiment and its results take place and support his findings. If the result disproves his theory, he without hesitation dutifully discards it or at least modifies it. In this way the body of identified knowledge is expanded, evolving in time to give us a more accurate picture of the world. Ah! If only it were like that.

Most scientists are formidably expert in their own speciality, are extremely knowledgeable in a wider region and, apart from hobbies, are often as ignorant in everything else as anyone else. One might expect however that their scientific training should give them some advantage in assessing the validity of anything new brought to their attention. Nevertheless scientists are human too and the modern generations are well aware that they live in a world of politicians, propaganda and spindoctors, a world awash with a torrent of ephemeral frothy and downright worthless media pap for people of limited attention, education and capacity for rational thought. In their own speciality scientists know what they and previous generations of researchers have found. Anything that drastically threatens to challenge the fortress of their hard-won and repeatedly tested and applied consensus of opinion is automatically suspect until supported by replicable experiment. Indeed, the greater the threat, the more reluctant the scientist will be to undertake the required experiments, especially if the person putting forward the new idea is not a respected colleague. In the past, many scientists have demonstrated a hostility to such challenges to seemingly well-established natural laws in their speciality. It is no good the aggrieved pioneer complaining that surely history has shown that the establishment has always spurned or neglected the maverick, the unconventional and the innovator only to accept his discoveries in the end. In this respect the wise words of Marx - not Karl but Groucho! - are relevant.

> 'They said Galileo was mad when he claimed the Earth revolved round the Sun – but it does. They said Wilbur and Orville Wright were out of their minds when they said men could fly – but they did. They said my uncle Waldorf was crazy – and he was as mad as a hatter!'

In modern times there are also the ever-pressing factors of time and money. In these days when funds for research are difficult to come by, there is enormous pressure from many quarters on the scientist to ensure that his available time is devoted to research projects that are 'respectable', grant-attractive and with promise of acceptable, immediately applicable results, improving the status and reputation of the institution that employs him or her.

In a real sense, all the above is relevant and comes within the science of celestial mechanics. But it is far more than that. In space research it also involves the design and control of the orbits and trajectories within the solar system of the spacecraft we launch together with the ability to know what it requires in rocket hardware to launch them. Some of its successes in this branch of celestial mechanics, called astrodynamics or astronautics, have been the placing in carefully tailored circum-Earth orbits of the hundreds of multi-purpose satellites for communication, Earth surveillance, observation of the far reaches of the universe; the missions to the Moon, Mercury, Venus, Mars, Jupiter, Saturn and beyond; missions to comets and asteroids; the Mariner and Voyager missions and the spectacularly successful Cassini-Huyghens mission to greatly enlarge our knowledge of Saturn and its system of satellites, particularly Titan. The astronautical dreams of Tsiolkovski, Hermann Oberth, Walter Hohmann and Werner von Braun became reality.

But wait a minute. Where in the book are all the elegant mathematical techniques in celestial mechanics such as Delaunay's lunar theory or Hamiltonian canonical equations, or general and special perturbation theories that have been developed over the past three and a quarter centuries since Newton's day, not only analytical but also computational especially after the invention and development of high speed and capacity electronic computers? They are an integral part, indeed a major part of celestial mechanics, in their detailed mathematical display a very beautiful story of the dedicated and tireless mathematicians who created them. And in Alessandra and Ettore's book these techniques are conspicuous by their absence. Certainly Newton's law of gravitation is given – twice, but where is everything else? Is this book rather like Hamlet with no mention of the prince? Is it in fact reminiscent of the story of the old Professor of Celestial Mechanics asked by an enthusiastic, mathematically inclined but celestial mechanics ignorant student to explain to him what celestial mechanics is all about.

> Professor, cautiously: *"You've heard of Newton's law of gravitation?"*
> Brightly: *"Yes!"*
> *"Well,"* even more cautiously, *"celestial mechanics is all about Newton's law of gravitation but you've got to know it very well."*
> Indeed you do.

I am of course being totally unfair to our authors. For they have planned their approach to their task carefully, they have hit upon the precise way in which it can be accomplished successfully, and in doing so they have identified correctly the readers they wish to capture and keep. Whether they be scientists already well-versed in mathematical and computational techniques, or historians of the march of science, or young high school students fascinated by the stream of astronomical information brought in by space research, or simply students interested in human nature, its motivations, its foibles, failures and triumphs, this book should please them. It is, in short, a good read.

The authors include accounts of many of the people who have contributed from the earliest times to our understanding of the phenomena of the heavens and the Earth's place under the celestial sphere. They relate how over four millennia ago in Mesopotamia careful records of eclipses of Sun and Moon, comets, and meteors were kept and attempts made to relate heavenly phenomena to terrestrial events such as famine and flood. We do not know the name of the person, possibly a priest in ancient Babylon, who, going through the astronomical library of clay tablets, discovered that very similar eclipses of the Moon occurred at intervals of 6585 days, a period of time named the Saros. The implications to him must have been staggering. If he could predict a heavenly event, what power it would give the priesthood in predicting terrestrial ones. Did he tell his fellow priests? Or did he, perhaps trembling with excitement, climb the steps of the ziggurat when the next lunar eclipse of that type was due? And only when he had witnessed it, did he reveal his discovery?

Subsequent pioneers such as Aratus, his poem *The Phaenomena,* Eudoxus and his sphere, Aristarchus of Samos, Eratosthenes, Hipparchus, discoverer of the precession of the equinoxes, Ptolemy and his epicyclic solar system, their theories and discoveries and the background to their lives are depicted. Later, Kepler, Galileo, Copernicus and Newton appear on the scene. Their lives, background and major contributions to astronomy are clearly given. Running through the accounts of their work is an important thread – motivation. These people were fascinated by heavenly phenomena, they wanted to understand why things happened, they got a 'fix' of supreme satisfaction when they believed their theories accounted for the phenomena. They sought the truth. In a real sense they became people we would recognise. They were scientists.

We read how the publication of Newton's *Principia* in 1687 – possibly only about twelve people fully understood its blockbuster nature when it appeared - opened the way to the following three and a quarter centuries of unprecedented progress in many scientific fields. And from then on in their book, Alessandra and Ettore skilfully lead us into the truly wonderful ongoing enterprise of astronomy in general and celestial mechanics in particular, the continuing display of new intriguing discoveries presented to us by our observational techniques and the responsive efforts by applied mathematicians to produce mathematical developments that could account for their observed dynamical features.

We learn of commensurabilities in mean motion, resonances, orbits described as horseshoes or tadpoles, with shepherd satellites guiding the particles of the ring between their orbits, other pairs of satellites that at their nearest proximity make a stately exchange of distances from their parent planet, dancing with intricate steps their pavane about the planetary maypole of gravitation. We learn of rings in profusion about Saturn, how unexpectedly crowded the outer region of the solar system is far beyond Pluto. We tackle the concept of stability of various kinds in the solar system and the attempts by celestial mechanicians to see if a chaotic system though unpredictable in the long run can still last for billions of years as if the system was indeed stable.

Mathematical celestial mechanics in fact finds its proper place in this book. Even without their elegant mathematics, the major contributions made by Newton, Lagrange, Poincaré, Hamilton and others are clearly described by the authors because of their own expertise in the subject and skill in presentation. Believing also that a picture is worth a thousand words, they have markedly increased the book's attraction by the choice, number and clarity of the diagrams and illustrations they include. And not the least of the book's value is the due attention they give to perhaps the hottest astronomical topic of the twenty-first century, the continuing discovery of planetary systems of other stars and the search for extraterrestrial life not only in our own Solar System but also elsewhere in the Universe. Giordano Bruno's heresy about the plurality of worlds will even receive its ultimate verification this century if life elsewhere is found. It would be a discovery equally as momentous in its implications for humanity as Darwin's theory of evolution.

I enjoyed this book. It is fresh and attractively written in its presentation of humanity's long-lasting love affair with the Universe, and I thank Alessandra and Ettore again for inviting me to provide the foreword.

Archie E. Roy
Professor Emeritus of Astronomy
Honorary Senior Research Fellow
Department of Physics and Astronomy
Glasgow University, Scotland

Authors' preface

Stability, resonances and chaos often sparkle as magic words in popular and scientific literature to explain the evolution of an astonishingly wide variety of complex systems, from weather forecasting to large-scale economies. Yet their origins can be traced back to an ancient discipline: celestial mechanics.

Born as a practical means to observe and predict the motion of the stars and planets, celestial mechanics has accompanied the history of any developing civilization on Earth. From the early astronomical observations of the ancient Chaldeans to the work of Henri Poincaré (whose intuition on the ubiquity of chaos continues to be an enlightening source of inspiration), until the Space Age, the number of celestial objects either discovered or launched into space has grown steadily. To date, more than 100,000 asteroids have been catalogued, the passages of thousands of comets have been recorded, satellites and rings are busily orbiting around the outer planets, and the still poorly known population of transneptunian objects extends far away into the outer reaches of the Solar System. This crowded Solar System has eventually forced astronomers to re-think the very same definition of a "planet".

The widespread diffusion of digital computers, the sharp increase in their performance, and significant advances in dynamical system theory, have allowed us to trace the orbital motion of celestial bodies over a timespan comparable to the age of our planetary system. On this timescale the Solar System is alive with events involving the major planets as well as the smaller bodies wandering among them.

On a larger scale, stellar systems and galaxies exhibit complex dynamical behaviour, while the long-awaited discoveries of extrasolar planets and their exotic orbital configurations are slowly bridging the gap between planetary science and astrophysics.

At the turn of the new millennium, humans have achieved routine access to near-Earth and interplanetary space. A cloud of artificial satellites for commercial, military and scientific purposes surrounds the Earth, and man-made celestial objects explore the Solar System, perform fundamental physics experiments and observe the Universe far from the disturbing presence of our planet. Their trajectories are confidently mastered by spaceflight dynamics. Stability, resonances and chaos therefore take us firmly back to celestial mechanics.

The idea of public education in celestial mechanics was somehow a logical consequence of our involvement in the organisation of the CELMEC meetings. Our aim was to gather together mathematicians, physicists, astronomers and engineers in order to facilitate communication amongst people working on celestial mechanics but associated with different institutions – universities, observatories, space agencies and industry. The enthusiastic international participation, the variety of topics discussed and their interrelations within apparently distant fields of study caught us by surprise. The chaotic behaviour of planetary spin-axes influences the long-term stability of climate on Earth which, in turn, has considerble implications for the birth of life on our planet. The orbital evolution of asteroids, comets and meteors implies a catastrophic impact causing the disappearance of the dinosaurs, and also puts into perspective the hazard for mankind. Spaceflight dynamics has quickly become a mature science, mapping the spaceways of the Solar System, and warning mankind against polluting the skies with orbiting debris.

We are aware that celestial mechanics has a longstanding reputation for being a rather complicated science; but an essential part of its fascination is that it has always been an ideal testing ground for the most complex mathematical theories. Bearing this in mind, we have tried to exploit as far as possible the graphic visualisation of the trajectories of the celestial bodies, thus minimising the use of analytical equations.

Historical highlights are frequently introduced for maintaining the reader's interest, as the circumstances of famous astronomical discoveries often follow intriguing plots typical of spy stories. Images have been also widely used throughout the text. Apart from the fascination of looking at alien worlds, the pictures returned by spacecraft have often shown the existence of unusual orbital configurations awaiting dynamical explanation.

This book was originally published under the title *Meccanica Celeste: il Valzer dei Pianeti* in 1996, and this translation is an updated and extended version of that text. Our hope is that we have succeeded in presenting the subject in a 'user-friendly' form to the non-scientist, as well as stimulating those more familiar with the technical aspects into making connections among the various fields of study which characterise the interdisciplinary nature of modern celestial mechanics.

Alessandra Celletti and Ettore Perozzi
Rome, 30 May 2006

Acknowledgements

Many people have helped us throughout the writing of this book: discussions, suggestions, comments, encouragement, logistic and moral support have been supplied by Niccolò Argentieri, Antonella Barucci, Teresa Boccuti, Ruggero Casacchia, Paola Celletti, Luigi Chierchia, Raffaele Chierchia, Carolina Ciampaglia, Grazia Ciminelli, Alessandro Coletta, Marcello Coradini, Simonetta Di Pippo, Elisabetta Dotto, Sylvio Ferraz-Mello, Lorenza Foschini, Giangiacomo Gandolfi, Margherita Hack, Corrado Lamberti, Edizioni Lapis, Cristina Lupi, Alessandro Manara, Paolo Marpicati, Barbara Martellacci, Andrea Milani, Luca Missori, Cristina Morciano, Franca Morgia, Jane O'Farrell, Anna Parisi, Walter Pecorella, Mirella Perali, Elena Perozzi, Giuditta Perozzi, Umberto Rampa, Enrico Romita, Alessandro Rossi, Hans Scholl, Bruno Sicardy, Bonnie Steves, Sabina & Simone Tonon, Giovanna Tranfo, Alberto Tuozzi, Giovanni Valsecchi, Giovanni Verardi. A very special thank to all the CELMEC participants, to Paolo Ulivi, who provided the missing link between Praxis and ourselves, and to John Mason and Clive Horwood for their patient and continuous support.

List of illustrations

1

Around and around

Your head is your house; furnish it.
Arabian proverb

The aim of modern celestial mechanics is to compute and predict the motion of celestial bodies, either natural or man-made. In ancient times it was part of the newly born astronomical sciences that exploited the knowledge of the regular movements of the constellations and the erratic motion of the planets (including lunar phenomena) for both religious beliefs and everyday life (such as the compilation of calendars). The concept of *orbit* was introduced after the development of cosmological models, and it dominated the evolution of celestial mechanics as a science. The Copernican revolution, the subsequent discovery by Kepler of the eccentricity of planetary orbits, and the Newtonian synthesis unveiling gravity, provided a model for the motion of celestial bodies which still holds today. Circles, ellipses, parabolas and hyperbolas - despite their simplicity from a geometrical point of view - represent the key to understanding the dynamical evolution of our Solar System and beyond.

THE IMPORTANCE OF BEING LUCRETIUS

The birth date of celestial mechanics is highly uncertain. Many ancient texts and writings have reached our epoch badly preserved or incomplete, and it is therefore not an easy task to trace at what period the study of the motion of celestial bodies merged astronomical observations and classical mechanics. Glimpses of past celestial mechanics appear in Roman and Greek literature, as in the masterpiece by the Latin poet Titus Lucretius Carus (98–55 BC), *De Rerum Natura* (*On the Nature of the Universe*, translated by Cyril Bailey, Oxford, 1947/1986), written to illustrate, in verse, the thoughts of the Greek philosopher Epicurus:

> The ship, in which we journey, is borne along,
> When it seems to be standing still;
> Another, which remains at anchor, is thought to be passing by.
> The hills and plains seem to be flying astern,
> Past which we are driving on our ship with skimming sail.

> All the stars, fast set in the vault of the firmament, seem to be still,
> And yet they are all in ceaseless motion,
> Inasmuch as they rise and return again to their distant settings,
> When they have traversed the heaven with their bright body.
> And in like manner Sun and Moon seem to abide in their places,
> Yet actual fact shows that they are borne on.
> And mountains rising up afar off from the middle of the waters,
> Between which there is a free wide issue for ships,
> Yet seem united to make a single island.

Describing the motion of ships at sea, Lucretius introduced the idea of relative motion – a crucial problem in physics. Due to modern means of transportation we know that it is possible to travel at very high speed without actually feeling it. If we do not see the landscape quickly passing by, or if we do not feel the bumps on the road, there is no way of detecting that we are in motion. In the seventeenth century Galileo Galilei called this the 'principle of inertia', by which any object, celestial or otherwise, has the tendency to maintain its state of rest or uniform motion (at constant speed along a straight line). If everything moves at the same speed, everything appears to be still. Three centuries later Albert Einstein extended this concept by saying that there is no such thing as 'rest', because motion is always relative to someone or something. And he further extended this concept. In his theory of General Relativity he investigated the very same nature of inertia, concluding that it is a consequence of the structure of the Universe. So, after a long journey in science and time we are back to the stars.

In subsequent verses Lucretius turns to the sky and focuses on the movements of the Moon, the Sun and the stars, remarking on the difference between apparent and true motion. The stars are 'fixed' because their relative positions, forming the familiar shapes of the constellations, do not change while the celestial sphere rotates as a whole over our heads at night. On the contrary, objects belonging to our Solar System 'revolve', and their position changes with respect to the background stars, even if at a rate much too slow to be caught by the naked eye. Yet to anyone who has enough patience to keep track of the position of the Sun, Moon and planets night after night, it is a compelling evidence. The resulting trajectories are not random, but have peculiar shapes and periodicities, thus indicating that some sort of regular motion is in action.

Finally, the illusions generated by perspective warn us not to trust our senses when dealing with astronomy, and distant objects and the rotation of the Earth may confuse the non-skilled observer. In the writings of Lucretius there is enough to believe that the study of the motion of terrestrial and celestial objects was already being merged two millennia ago.

Although the visionary Greek astronomer Aristarchus (310–230 BC) had already conceived an heliocentric Solar System, the belief that the Earth was located at the very centre of the Universe dominated western astronomy for almost 2,000 years. There are many reasons for such an enduring belief. Firstly, the assumption that the Sun and the other celestial bodies are revolving around the Earth is the model that is by far closer to the 'ground truth' of everyday experience. Our twenty-first-century educated minds have been taught that our planet moves along its orbit at an astounding velocity of around 100,000 km/h while spinning around its own axis and carrying the entire biosphere. But our senses do not perceive any motion. Modern physical concepts must be utilised to explain why we should not believe our senses, and exact experiments need to be performed in order to detect and measure the motion of our planet.

The geocentric system assessed by Aristotle (384–321 BC) was officially adopted at the beginning of the second century AD, by Ptolemy in his *Almagest* – a work in thirteen books, the aim of which was to summarise the astronomical knowledge of that time. In the following centuries the Aristotelian–Ptolemaic cosmology was strongly supported by the Christian Church as proof of the human excellence within God's creation. Thus 'geocentrism' represented the perfect mix of science and religion – a reasonable view of the Universe, with Man placed at the very centre and the Holy presence manifested by the perfect circles drawn by the celestial bodies in the skies (Figure 1.1).

Yet when more and more accurate measurements of the motion of Solar

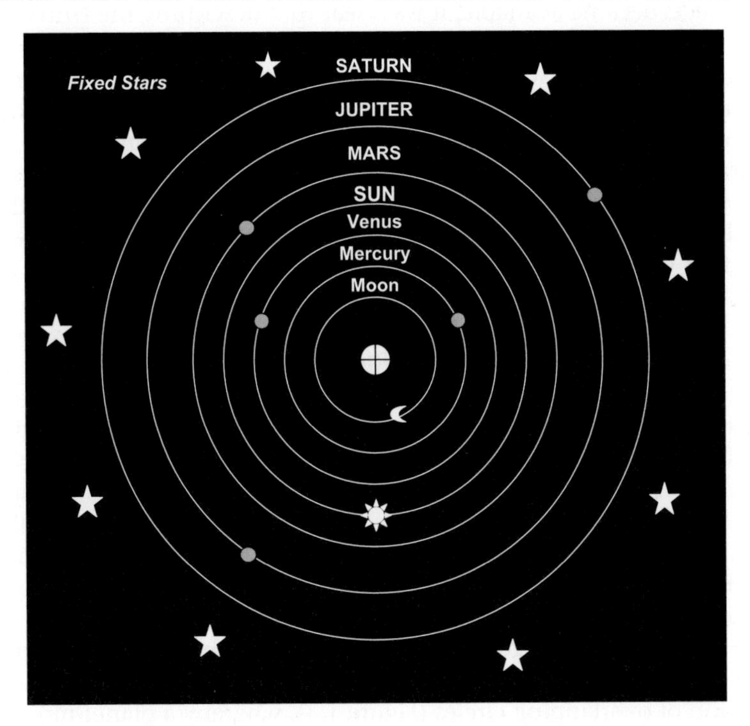

FIGURE 1.1. The geocentric system. The Sun, the Moon and the five naked-eye planets revolve around the Earth. No reliable estimate on their size or distance is provided.

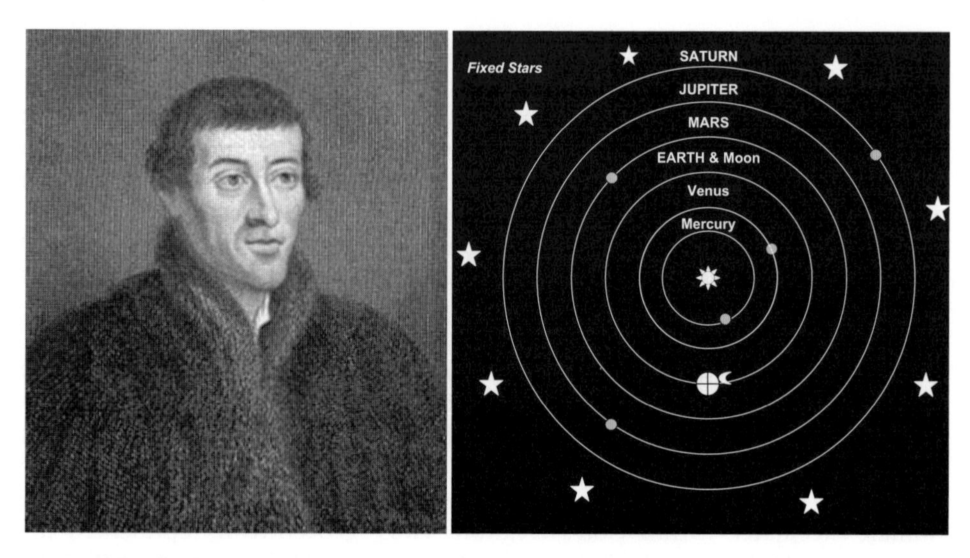

FIGURE 1.2. Nicolaus Copernicus and the heliocentric system described in his book *De revolutionibus orbium coelestium* (*On the Revolutions of the Heavenly Spheres*).

System objects became available, it was clear that describing the Universe as seen from an orbiting 'spinning top' was fitting much better astronomical observations. The geocentric system was eventually discarded at the beginning of the sixteenth century, when Nicolaus Copernicus, after much hesitation due to the dangerous political and religious situation (the Lutheran schism had just occurred), published his six-book work *De Revolutionibus* (*Orbital Revolutions*). Herein he dismissed the Earth from the centre of the Universe and introduced a new model in which all planets revolve around the Sun (Figure 1.2). It was 1543 – the year that Copernicus passed away.

ECCENTRIC KEPLER

In the early 1600s the astronomer and mathematician Johannes Kepler (1571–1630) introduced the modern definition of orbit. The Latin word *orbit* means 'circle', and before Kepler, scientists – Copernicus among them – assumed that celestial bodies moved with constant velocities along strictly circular paths, wherever the centre of motion. But the observed positions of the planets indicated that their velocity is not constant in time, instead showing periodic accelerations and delays.

At that time it was dangerous to deny circular orbits, and a rather intricate modelling was therefore proposed without having to do so. This involved the construction of overlapping circles (Figure 1.3), whereby a planet moves steadily along a minor circle – the *epicycle* – the centre of which advances at constant speed along a circular orbit around the Sun – the *deferent*. The composition of the

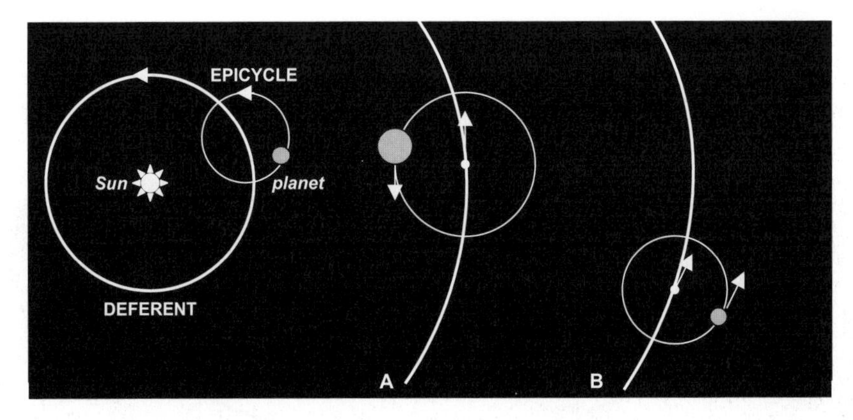

FIGURE 1.3. If a planet moves on a small circle whose centre follows an heliocentric orbit it travels slower when the two velocities act in opposite directions (A) and faster when the velocities adds up (B).

two uniform motions along circular paths could explain the observed velocity variations. But the more accurate the observations, the more 'epicycles' needed to be added for the model to be reliable. Circles of ever decreasing size circled around circles, thus producing an increasingly complex description of the motion of the planets.

Kepler broke the circular paradigm and showed that a simpler and more accurate model of the Solar System could be produced by using elliptical orbits, and that the motion of the planets is ruled by a few quantitative relationships.

Kepler's laws

Kepler's three laws of planetary motion are stated as follows:

First law	The orbit of each planet is an ellipse with the Sun at one focus.
Second law	The radius vector joining planet to Sun sweeps out equal areas during equal time intervals.
Third law	The cubes of the semimajor axes of the planetary orbits are proportional to the squares of the corresponding periods of revolution.

There are a number of easily understandable consequences of these laws. According to the first law, the Sun is located in one of the foci of the ellipse (Figure 1.4); therefore, when a planet undergoes a complete revolution it has a minimum and a maximum distance from the Sun. These points are called *perihelion* and *aphelion*, and are usually indicated by the letters q and Q respectively (Figure 1.4). The larger the difference between perihelion and aphelion, the more eccentric the orbit: as indicated by the parameter $e = (Q-q)/(Q+q)$, which is called 'eccentricity' of the ellipse.

The remaining two laws rule the motion of celestial bodies on their orbits. The distribution of velocities along an elliptic orbit reaches a minimum at aphelion,

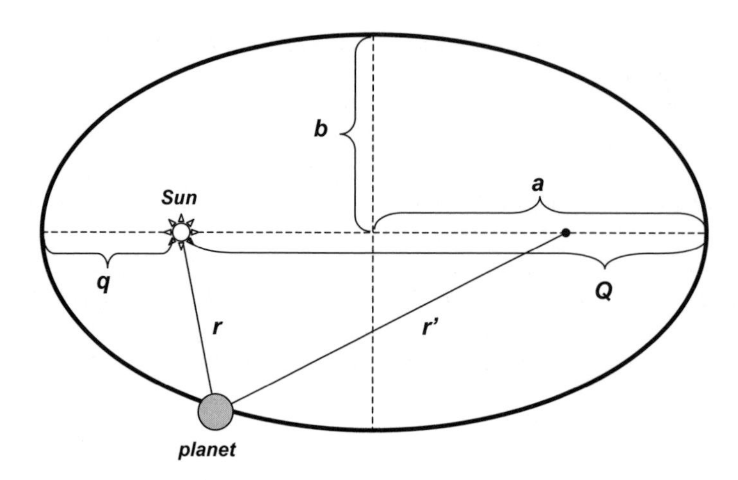

FIGURE 1.4. Graphical representation of the Keplerian orbital parameters. The Sun is located at one of the two foci of the ellipse, which is drawn by keeping constant the sum of the distances from the foci ($r + r' = 2a$).

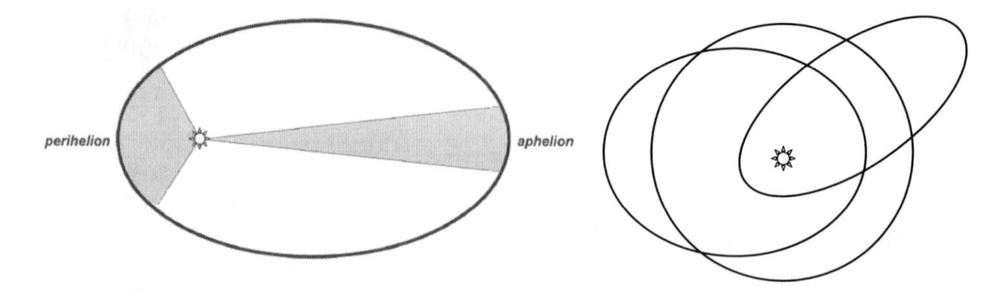

FIGURE 1.5. (Left) Graphical representation of Kepler's second law. At perihelion a larger branch of ellipse is travelled in the same timespan, thus implying a higher orbital velocity. A consequence of the third law is that the orbits at right, having the same semimajor axis, also have the same revolution period.

and is maximum at perihelion. The comparison between the areas mentioned in the second law (Figure 1.5) allows an estimate of these variations. In a highly eccentric orbit a celestial body spends most of the time far from the centre of motion, near aphelion, allowing only quick apparitions close to the Sun. Some comets are observed to move on highly eccentric orbits.

Kepler's third law implies that the period of revolution does not depend on the shape (precisely on the eccentricity) of the orbit, but solely upon its mean distance from the Sun, expressed by the value of the semimajor axis which measures the distance from the centre of the ellipse to the aphelion (equivalently, to the perihelion). Orbits with identical semimajor axes are therefore completed within the same timespan, whatever the eccentricity (Figure 1.5). It will be helpful to bear this in mind when discussing *mean motion resonances* – a typical behaviour which affects some bodies of the Solar System (as extensively discussed in Chapter 3).

Kepler's laws were originally stated to describe the orbital paths of the planets, but they can be applied to any system of celestial bodies such as the motion of natural or artificial satellites, with a modification in terminology. For Earth-orbiting bodies it is *perigee* and *apogee*; for the Moon, *periselenium* and *aposelenium*; and so on (the generic form is 'apocentre' and 'pericentre'). The term 'Keplerian motion' also has a wider meaning, and is not restricted to elliptical orbits. From a geometrical point of view it is possible to show that there are three basic orbital shapes, corresponding to an *ellipse,* an *hyperbola* and a *parabola.* These curves are known in elementary geometry as *conic sections* (obtained as intersections of a plane with a cone), and the motion of celestial bodies along them is ruled by Kepler's laws. In particular, hyperbolic paths are typical of high-velocity close encounters between celestial bodies, thus also providing the basic description of a *gravity assist* – a technique widely used for interplanetary probes to obtain 'free' momentum from the gravitational field of a planet.

Conic sections

An ellipse (Figure 1.4) is a closed curve with a size and shape determined by the values of two characteristic quantities: the *semiaxes* of the ellipse. The longer is called the *semimajor axis,* indicated using letter a, while the shorter is b.

The two points lying on the major axis are the *foci* of the ellipse. These have the peculiar property that the sum of their distances from any point on the ellipse is constant and equal to $2a$. A measure of the flattening of the ellipse is given by its *eccentricity* e, the value of which ranges between 0 and 1, and it is related to the different length of the semiaxes (Figure 1.6). When $a = b$ then $e = 0$ and the ellipse is a circle; and as eccentricity increases the ellipse becomes more and more elongated until when $e = 1$ it degenerates into an open curve: a *parabola.* If the eccentricity is greater than 1, an *hyperbola* is produced (Figure 1.7).

Parabolic motion somehow represents an abstraction, since the corresponding conic section is the one separating ellipses from hyperbolas. Yet it is historically associated with a certain class of comets – those coming from remote regions of the Solar System and entering the planetary region on orbits so elongated that it is difficult to decide whether they have a slightly hyperbolic or an extremely elliptical shape.

Returning to the planets: in Kepler's view, Mercury, Venus, Earth, Mars, Jupiter, Saturn, Uranus and Neptune (Pluto, as it will be explained later, is a remarkable exception) revolve along almost circular orbits at increasing distances from the Sun. The separation between them is wide enough and their eccentricities small enough to avoid intersecting each other – a good point in favour of the stability of the Solar System as a whole. But the orbits of the planets

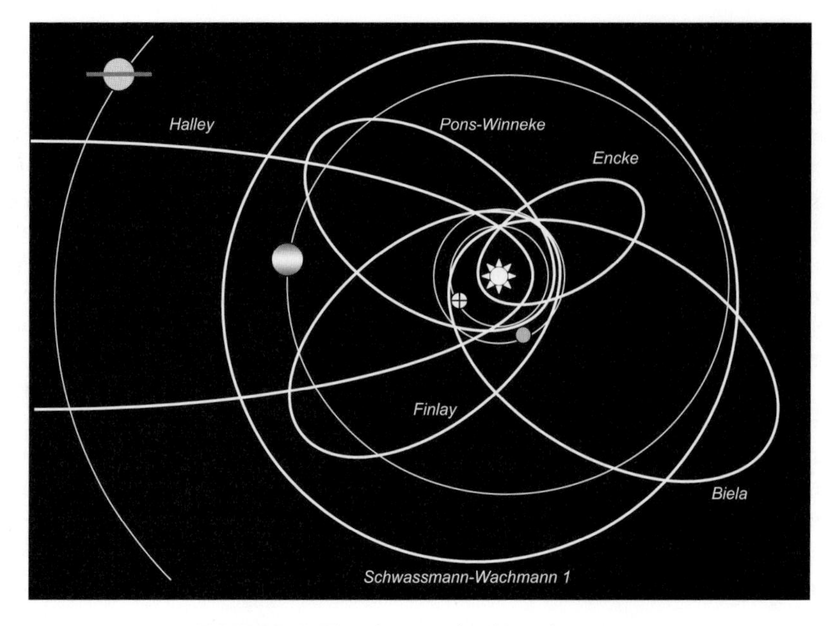

FIGURE 1.6. The elongated orbits of comets.

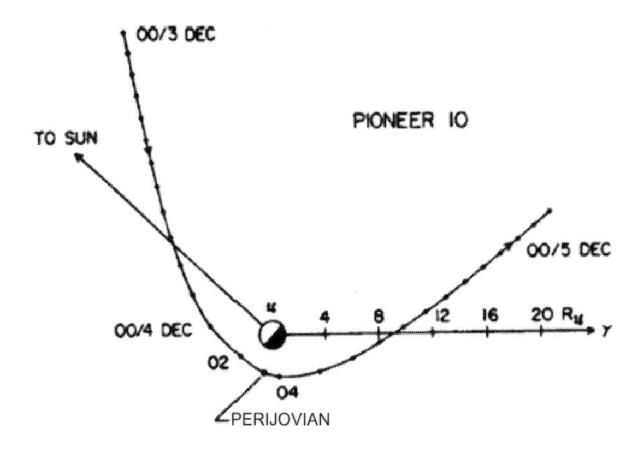

FIGURE 1.7. The hyperbolic trajectory of Pioneer 10 at its encounter with Jupiter in 1973. (Courtesy NASA.)

possess another important property: they are almost coplanar, in the sense that the planes containing the orbital ellipses are not significantly inclined with respect to each other. But what happens when the orbits are not so tidy, as in the case of the near-Earth asteroid population or as it is observed in some exotic planetary systems discovered around other stars? The whole picture becomes more complicated, because ellipses must be now imagined free, to be reoriented in three-dimensional space (Figure 1.8). The only fixed point is their common focus where the central body (such as the Sun) resides. As an example: by simply

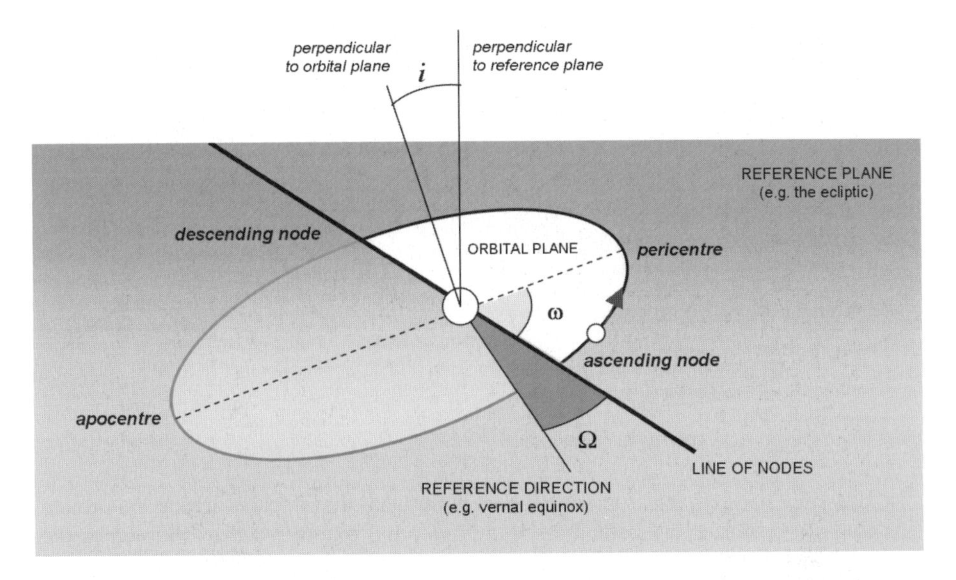

FIGURE 1.8. The orientation of an orbit in space is defined by three angular parameters: the inclination *i* with respect to a reference plane; the intersection of the orbital plane with the reference plane – the *line of nodes* – the direction of which is defined by the longitude of the nodes, Ω; and the angular distance of the pericentre from the line of nodes, measured by the argument of pericentre, ω.

rotating two identical ellipses in their own planes one obtains completely different relative geometries. The two orbits can be locked one inside the other without ever touching, like two subsequent elements of a chain, or they can intersect each other at the ascending or descending nodes (Figure 1.9).

While reviewing the long list of open problems in celestial mechanics, Archie E. Roy, of the University of Glasgow, once said that a more appropriate name for people working on this discipline is 'celestial plumbers'. If this is so, ellipses, parabolas and hyperbolas are the workman's tools for fixing leaks in the celestial sphere.

TYGE THE ASTROLOGER

In his book *A History of Astronomy*, Anton Pannekoek depicts Kepler as one of the first modern men of science, oriented toward a new approach essentially based on the search for connections between cause and effect of physical phenomena. Using Pannekoek's words: 'The previous generation had asked of any phenomenon: What does it mean? The new generation asked: What is it and what is its cause?'

This statement is undoubtedly true when considering Kepler as opposed to his teacher, Tycho Brahe (1546–1601). Nevertheless, Brahe's refined observations of the position of the planets (particularly Mars) played a fundamental role in

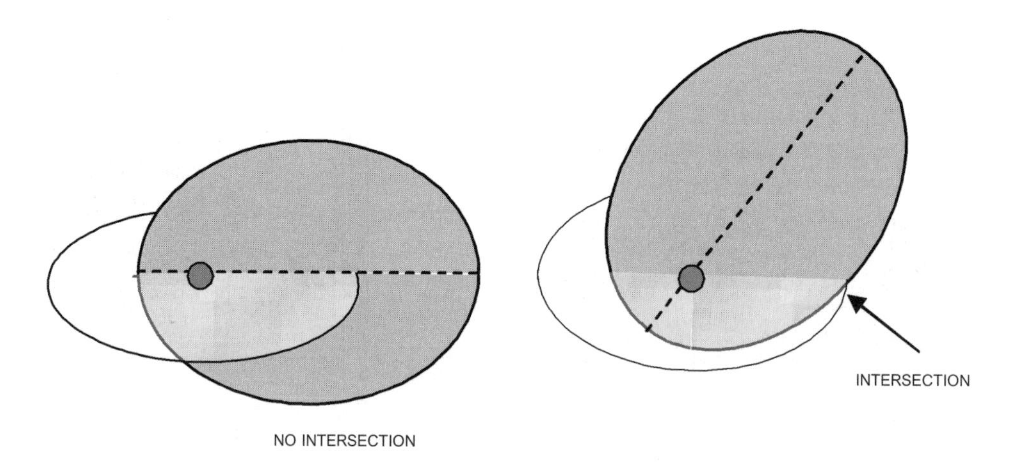

FIGURE 1.9. Changing the argument of pericentre ω has the effect of rotating the orbit in its own plane, which may result in bringing two orbits to intersection, as shown by the arrow. This behaviour is typical of the long-term dynamical evolution of the orbits of near-Earth asteroids, and raises the risk of collisions with the Earth.

allowing Kepler to place the planets on their elliptical orbits. Yet it does not tell the whole story.

The Danish nobleman Tyge Brahe (Tycho is the Latinised name) was born in 1546. Being strongly impressed by the eclipse of 21 August 1560, he soon abandoned juridical studies to dedicate his life to observing the sky and to predicting its phenomena. He became well known in Europe for his discovery, on 11 November 1572, of a spectacular *nova* – a 'new star' in the constellation Cassiopeia – that was so bright that it was visible even in daytime, after which it faded and disappeared after almost two years. The studies carried out by Tycho on this celestial phenomenon led to his being introduced to King Frederick of Denmark – an enlightened man interested in science. The King presented Tycho with the island of Hven, where in 1576 he built the first astronomical observatory, called Uraniborg (Figure 1.10). Accurate predictions of the motions of the planets soon became Tycho's main interest, and he was firmly convinced of the influence of the heavenly bodies on Earthly events, such as wars and pestilences. His predictions required sufficient accuracy, which was not always possible. For example, the predicted conjunction of Jupiter and Saturn in 1563 had displayed an error of one month! To overcome this hindrance, at Uraniborg Tycho designed and built innovative instruments to improve the accuracy of the determination of planetary positions. He soon began to perform regular observations of the planets, but in 1588 King Frederick died – and his successors did not share his interest in science. Tycho was then forced to leave Denmark, and in 1599 he arrived in Prague, taking with him his precious data. There he met the young Kepler, to whom he denied access to his records of planetary positions. This is not surprising, as for an astrologer such data represent the real

FIGURE 1.10. Astronomical observations at Uraniborg.

knowledge, and Tycho wanted to interpret the observations himself. However, this was not to be, and two years later he died in mysterious circumstances. Murder by poison cannot be excluded; but whatever the truth it can be said that after the astrologer Tyge it was 'written in the stars' that someone else should have profited by his work.

GIANTS' SHOULDERS

On 5 February 1675 Isaac Newton (1643–1727) wrote to Robert Hooke: 'If I have seen further than others, it is by standing on the shoulders of giants'.

Kepler was one of the giants. Let us climb on his shoulders and peer beyond the third law, relating the size of an orbit (the mean distance from the Sun of a celestial body, measured by the semimajor axis a) to the period of revolution. This law implies that the time needed to complete an orbit around the Sun increases as we move farther from our star. Although it might seem obvious, because ellipses increase in size, the mathematical relationship involved in Kepler's third law dictates that as we recede from the Sun, trajectories are also covered at a *slower* speed. As an example we compare the motion of our planet to that of Jupiter. The length of the Earth's almost circular orbit amounts to about 1 billion km, which are covered in 365 days. It is easy to see that our planet has a mean motion of nearly 1° per day, corresponding – using more familiar units – to an orbital velocity of 2.5 million km per day, or 30 km/s. Jupiter is five times farther from the Sun than the Earth, and its orbit is five times longer. Therefore if Jupiter were moving at the same speed as the Earth, its period of revolution would be about five years. However, Jupiter completes its orbit in almost 12 years. If there is something that drives the planets along their orbits, its influence becomes less and less as we recede from the centre of motion. Newton called that 'something' *universal gravitation*, and provided a detailed description of the phenomenon.

Gravity

Two bodies of mass M_1 and M_2 attract each other with a force F stronger for larger masses and weakening when the mutual distance d increases:

$$F = -G \frac{M_1 M_2}{d^2}$$

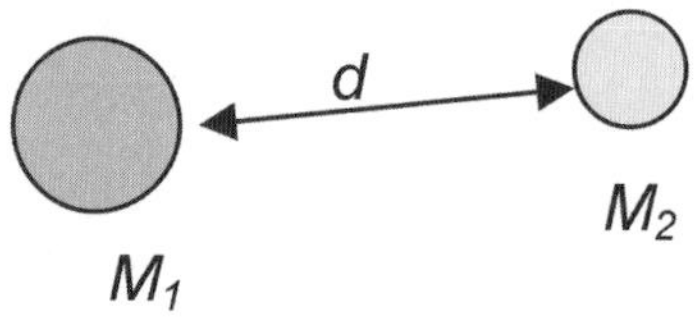

d

M_2

M_1

The negative sign indicates that the force is attractive, while G is the *gravitational constant*. Despite its simple form, this relationship accounts for the motion of all celestial bodies, however complex their trajectories. The equations of motion for a system composed by N bodies is obtained by simply adding the gravitational terms of every possible 'couple'. If the masses M_1 and M_2 are that of the Sun and of a planet, it is possible to deduce Kepler's laws. Conic sections are therefore referred to as the solution of a 'two-body problem' – an ideal system composed of only two celestial bodies. As will be explained in detail in the next section, the reason why planetary orbits are so close to two-body solutions is that their masses are relatively small when compared to that of the Sun, and that they are sufficiently apart to minimise the direct attraction of one planet with the others.

The formulation of the laws of gravity brought an enlightening unified vision into the scientific knowledge of that time. One single physical law could explain a number of apparently different natural phenomena: the orbital paths of the planets, the Earth's attraction over falling bodies, and the origin of lunar tides. Newton also developed the necessary mathematical tools for computing the behaviour of gravitationally interacting bodies, now widely known as *differential calculus* (although modern calculus derives from Leibnitz). It soon became a powerful tool for astronomers, and the study of the dynamics of Solar System bodies governed by gravitation grew as a self-standing discipline. The French mathematician Pierre Simon de Laplace (1749–1827) named it 'celestial mechanics'.

PERTURBATIONS

The Sun has a mass about 1,000 times larger than the total mass of all the planets in the Solar System. Thus, according to gravitation, our star is the 'engine' driving planetary motion. Satellites are also observed to orbit the planets: large and alone like the Moon, or in good company as with Jupiter (more than sixty satellites to date) and Saturn (almost sixty satellites). Many other celestial objects travel through interplanetary space. Icy comets are everywhere in the Solar System, while small and irregularly shaped asteroids are mostly confined between the orbits of Mars and Jupiter. Some of them - the NEAs (near-Earth asteroids) - are often approaching the terrestrial planets on chaotic orbits. The first member of the long-sought population of transneptunian objects (TNOs), located at the border of the planetary region, was found in 1992, and to date their number approaches 1,000.

Any attempt to apply Newton's gravitation to this crowded ensemble of celestial objects immediately presents problems. All these bodies, no matter what their size, have non-zero mass, and gravity is present in each of them ('no exceptions', Newton used to say). Thus the Earth feels the attraction of the Sun, but it is simultaneously attracted - even if to a minor extent - by the gravitational pull of the other celestial bodies. The direct consequence is that Kepler's laws hold exactly only if gravitation acts independently for each two-body subsystem's Sun–planet. But this is not the real case, and celestial mechanics has been forced to review the concept of orbit with the introduction of the *perturbations*.

Perturbing the Solar System

Imagine turning off gravitation among the planets, leaving only the reciprocal attraction between the Sun and each individual planet. In this case the motion of the nine planets is described by a set of nine independent Newtonian equations, each involving the mass and the distance which characterise the specific Sun–planet pair. The resulting orbits are ellipses of increasing size.

Now imagine switching on the gravitational attraction by one of the planets. This implies that another body acts gravitationally within the system, and it is necessary to include an additional 'Newtonian' term to every planetary equation. For example, if only Jupiter is gravitationally active, one needs to add to the equation of motion of the Earth a term of the same form as Newton's law in which M_1 is the mass of Jupiter, M_2 is the mass of the Earth, and d is the Earth–Jupiter distance. The orbit of the Earth is then said to be *perturbed* by Jupiter. Such terminology indicates that the trajectory of the Earth exhibits only small variations under the effect of Jupiter, since its mass is much smaller than that of the Sun, which always plays the role of director of the Earth's orbital motion. This procedure can be repeated until all the bodies of the system are restored into full action and their individual perturbations are computed.

In our Solar System the Sun's gravity is considerably stronger than any planet-to-planet interaction. Perturbations among the planets are small enough to allow planetary orbits to closely resemble ellipses, for the sake of Kepler's laws.

Distant planetary perturbations result in tiny deviations from the Keplerian motion, and can be visualised as small changes in the elliptical shape of the orbit. If some orbital element such as the eccentricity or the semimajor axis oscillates around some given mean value, then the orbit experiences periodic contractions and expansions, behaving like a string slowly vibrating around its position at rest. A well-known example of this phenomenon is the Great Inequality, involving the two most massive planets of the Solar System, Jupiter and Saturn, with orbits that oscillate with a period of about 900 years. When no periodicities are involved, but the orbit changes steadily in time, then a *secular perturbation* is in action. The rotation of an orbit in its own plane is a typical example of this kind. The consequence is that the perihelion occurs at slowly drifting directions, thus changing the geometrical relationship with other orbits. Secular perturbations are responsible for bringing two orbits to intersection, thus providing an evolutionary link between the two situations described in Figure 1.9. In extreme cases – such as during the close encounter of a comet with a giant planet – the orbit may be so perturbed that it cannot be reduced to a simple geometrical shape (Figure 1.11).

As new celestial objects were discovered exhibiting complex and unexpected dynamical evolutions, celestial mechanics classified the corresponding orbital motions and developed sophisticated mathematical techniques to deal with all different types of perturbation. Ellipses, parabolas and hyperbolas are said to be the solutions of a two-body problem, because they describe an orbital motion accounting only for the gravitational attraction of two bodies. Although it

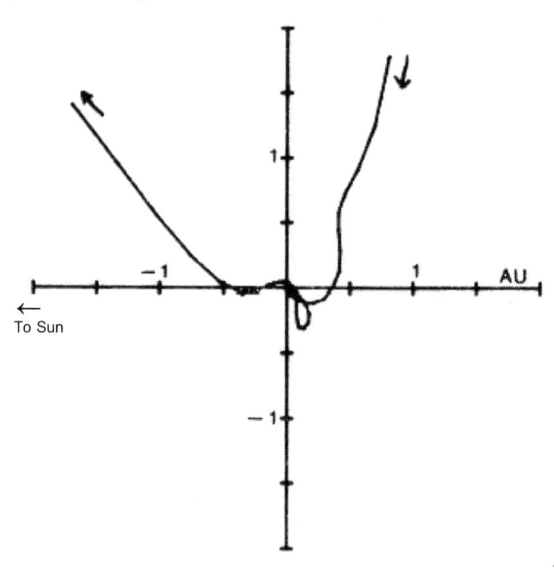

FIGURE 1.11. The orbital path followed by periodic comet Oterma during its close encounter with Jupiter during 1934–39, in a frame rotating with the planet. The comet became a temporary satellite of the planet.

represents a first-order approximation to reality, the two-body problem is a fundamental reference for studying more crowded systems. In the three-body and more-body problems a Keplerian orbit should be considered only as a snapshot of the real motion – the trajectory that a celestial body follows if the perturbations of other bodies suddenly disappear (the *osculating* orbit). Actual trajectories undergo steady changes, passing more or less dramatically from one Keplerian orbit to another according to the way in which perturbations act within the system. As we shall see in the following chapters, the resulting orbits might become so eccentric as to extend throughout the entire Solar System or tangle up in knots. This is when chaos enters the scene.

DRAWING ORBITS

Until now we have followed the great debate on the 'cosmological' definition of orbits and on the discoveries that led to the Newtonian theory of gravitation. Kepler's laws and perturbations can be translated into mathematical relationships and equations which allow the prediction of the position of a celestial body for a certain timespan. This machinery is usually referred to as *orbit propagation*, and in the pre-digital computer era it produced long lists reporting the time evolution of the position of celestial bodies on the celestial sphere. These tables – called *ephemerides* – are of fundamental importance, because they tell astronomers where to point their telescopes. After all, only the Sun, the Moon, the five planets from Mercury to Saturn, and some bright comets, are naked-eye Solar System objects!

When a new celestial body is discovered, the computation of its ephemerides is even more important because it allows the recovery of the object on subsequent nights, thus providing the necessary confirmation. The process of drawing a consistent orbital path that encompasses the observed positions in the sky is called *orbit determination* – which is not always an easy task (Figure 1.12). As an example: when in January 1801 Ceres, the first asteroid (now 'upgraded' to dwarf planet) was recognised by Giuseppe Piazzi (1746–1826) as a faint object moving among the stars, he could measure its position for only a few nights before it faded into the sunset background luminosity. It was several months before Ceres could be observed again, and by then the astronomers realised that the orbit was not sufficiently accurate to allow recovery. Ceres was lost, and after many attempts to find it again, some even began to doubt Piazzi's discovery. (This was many years before the invention of photography, and one had to trust the astronomer's word and eyes!) Luckily enough, the problem of Ceres' recovery was solved by the brilliant mind of the young Karl Friedrich Gauss (1777–1835) – one of the greatest scientists of all time. Gauss developed a new method for orbit determination that led to the recovery of Ceres on 31 December 1801, almost one year after its discovery. (The basic concept, called 'least squares analysis', is widely used in many mathematical and physical contexts.)

The reason why orbit determination can be problematic is that telescopic

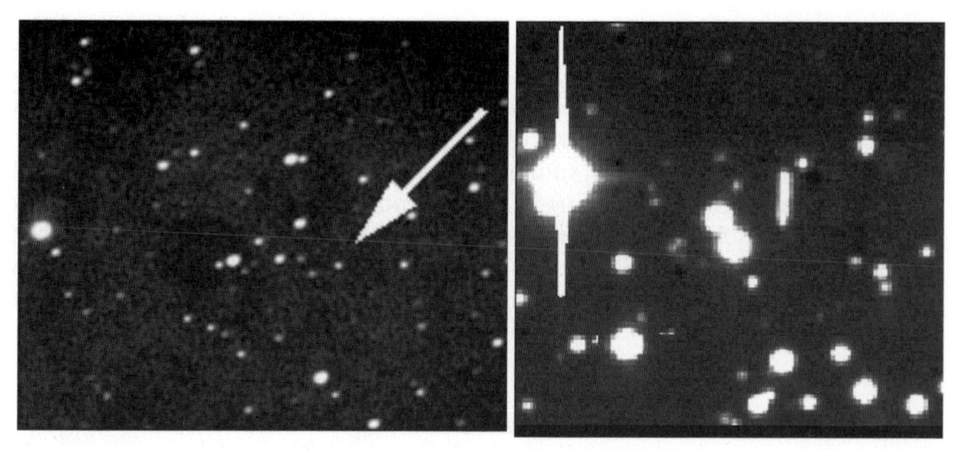

FIGURE 1.12. An asteroid (indicated by an arrow in the image at left) can be easily confused with the background stars, and in order to recognise its nature the apparent motion must be detected. The image at right is a long-exposure photographic plate on which a small bright track appears due to the asteroid's changing position in the sky. (The one shown is asteroid 7329 Bettadotto)

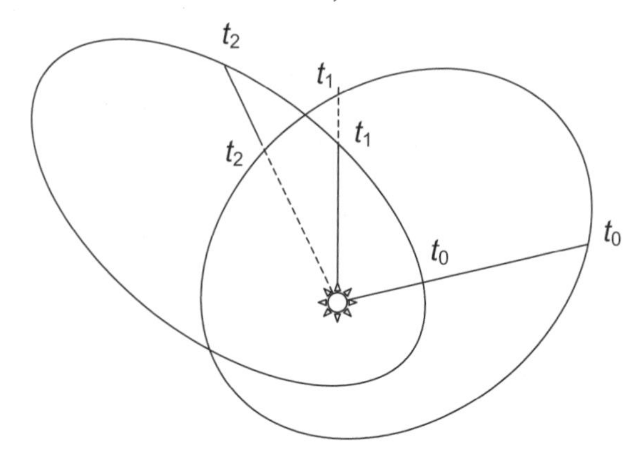

FIGURE 1.13. Without information on the radial velocity of an object as seen from the ground, observations too close in time t_1 and t_2 are compatible with widely different orbits. The finding of a prediscovery observation (t_0) is of considerable help in improving the orbit determination process, because the time needed to travel from t_0 to t_1 depends strongly on the shape of the orbit. (Relative geometries are exaggerated in the figure)

observations reveal only the apparent motion of an object on the celestial sphere. No information is provided on its radial motion – whether it is moving closer or farther along the line of sight of the observer. Moreover, the luminosity of a distant and/or small celestial body, especially at discovery, is usually close to the limit of the resolving power of the telescope, which translates into a large uncertainty in the position. The situation is depicted in Figure 1.13, which shows that completely different orbits can be drawn, both fitting the two observed positions on the celestial sphere. The more and more spaced in time are the

FIGURE 1.14. The astronomers involved in the discovery of Neptune: (left–right) John Couch Adams, George Biddell Airy, James Challis, Johann Galle and Urbain J.J. Leverrier.

observations, the more accurate the orbit determination. This is the reason why when a new discovery is announced, word is quickly spread among astronomers to concentrate the efforts of observing the new celestial body. Yet, as we have seen in Piazzi's case, it is not always possible to do so.

A significant improvement can be achieved by searching prediscovery observations in the archives of observatories around the world in order to determine whether the new celestial body was previously observed but erroneously considered as a background star (Figures 1.12 and 1.13). It is a difficult and lengthy job, only recently alleviated by the use of modern computerised automatic search techniques. It is, however, a highly successful tool, because it frequently happens that, while focusing on the primary goal of an astronomical observation (a comet, a galaxy, and so on), only marginal attention is paid to the background stars.

The most famous case of prediscovery is that of Neptune (Figure 1.14). Its position was already noted by Galileo during his observations of the jovian satellites in 1612; again, in 1795 the French astronomer Joseph-Jerome Lalande catalogued it as an 8th-magnitude star; and finally, in September 1846 - only a few days before the official discovery of the planet - it was in the field of view of the telescope of Johann Von Lamont, at the Berghausen Royal Observatory, near Münich. In this respect Neptune is a planet whose destiny seems always to put astronomers in big trouble. For example, in recent years ground-based observations suggested that the planet was surrounded by 'arcs' instead of a regular ring system; and the puzzled astronomical community had to wait until the Voyager 2 spacecraft reached the planet in 1989 to solve the riddle. But there is no doubt that the events which led to the discovery of Neptune are among the most intriguing and breathtaking in the history of astronomy - a complex interplay of scientific skills, intuition, human factors and the significant role that good and bad luck may still play in science. It is worthwhile recalling the details, because all fundamental methods of celestial mechanics discussed so far are involved: Keplerian motion, perturbation theories and orbit determination.

NEPTUNE'S KARMA

Towards the beginning of the nineteenth century the steady improvement of observational techniques enabled measurement of the deviation of planetary orbits from exact elliptical shapes with a high degree of precision. At the same time, celestial mechanics developed refined planetary theories in order to take into account perturbations and to compute accurate ephemerides. On 13 March 1781 William Herschel had discovered Uranus – the first planet found with the aid of a telescope. With its orbit located beyond Saturn, the possibility that other unseen more distant planets awaited discovery fascinated astronomers. The Solar System seemed to be much larger than previously thought. The most promising clue in this direction was the existence of unexplained discrepancies between the computed and the observed orbit of Uranus. The attention of the astronomical community was concentrated at this point, as witnessed by the words of the Director of the Paris Observatory, Alexis Bouvard, who in 1721 wrote: 'I leave it to the future the task of discovering whether the difficulty of reconciling [the irregularities found in Uranus's motion] is connected with the ancient observations or whether it depends on some foreign and unperceived cause which may have been acting upon the planet'.

No wonder, then, that in 1845 the young British astronomer John Couch Adams (1819–1892) was busily developing an innovative technique to reverse the classical problem of celestial mechanics. Instead of computing the orbital evolution of a planet taking into account perturbations by another planet, he tried to find the position of a planet by considering the perturbations exerted on another planet. Towards the end of 1845 he wrote to the Astronomer Royal, Sir George Biddell Airy, that he had produced a reasonable prediction of the position in the sky of the unknown perturber of Uranus, and urged observations. Unfortunately Airy did not trust his results – possibly because Adams was quite young; but he missed the point, arguing instead on some issues of minor importance. As a consequence, no observations were performed in the region indicated by Adams. Some months later, in June 1846, the French astronomer Urbain Jean Joseph Leverrier (1811–1877) published his results on the same problem – and they were in good agreement with those of Adams. The difference between the two predictions in the position of the unknown planet was less than 4°. Upon seeing this remarkable agreement, Airy alerted the Director of Cambridge Observatory, James Challis, who immediately began a survey in that region of the sky. But he failed to reduce his observations quickly enough, and on the night of 23 September 1846 Neptune was found from the Berlin Observatory by Johann Gottfried Galle, who that same evening had received a letter from Leverrier. His answer was prompt: 'Monsieur, the planet of which you indicated the position really exists'.

The discovery of Neptune was welcomed worldwide as a striking example of the power of science over superstition and false beliefs. For astronomy it represented the triumph of celestial mechanics as a science, and perturbations played a key role in it – not only indicating the existence of an unknown planet, but also being used to predict its position in the sky. It was the first time that a celestial object was not found by accident or as the result of extended sky surveys, but rather directly from mathematical computations. Despite his bad luck, Adams' contribution was also widely acknowledged, and in 1854 Leverrier was appointed Director of Paris Observatory.

But this was not the end of the story. As the number of observations of Neptune's motion grew, the orbit of the planet became more and more reliable, leading to an unexpected result. Both Adams and Leverrier had assumed (following Bode's law, which will be discussed in Chapter 5) that the unknown planet had a mean distance from the Sun equal to 38.8 Astronomical Units (AU – the mean distance of the Earth from the Sun); but orbit determination was instead indicating that the planet was located at a mere 30 AU, and the semimajor axis of Neptune's orbit was more than 1 billion km different from the predictions. How could an incorrect hypothesis have led to correct results? In order to answer this question the American astronomers Benjamin Peirce (1809–1880) and Sears Cook Walker (1805–1853) carried out extended and careful investigations. Eventually they showed that between 1790 and 1850 Uranus and Neptune passed through conjunction (when two planets on the same side of the Sun are aligned with the Sun), and this had helped minimise the error in Leverrier's and Adams' initial guess of the semimajor axis. So, if the events had taken place at a different epoch (before or after), Neptune would have been much further from the predicted position in the sky, and as a consequence it would have been more difficult to find.

Peirce and Walker did not want to discredit the work of Leverrier and Adams, whose methods are still considered a major achievement for celestial mechanics. They simply added a touch of good luck to the discovery of Neptune. Unfortunately, Leverrier lacked positive thinking and spent many years trying to demonstrate that the conclusions drawn by the two American astronomers were wrong. He grew bitter and unfriendly, to the point that in 1870 he was removed from his position as Director of Paris Observatory.

The Neptune files

With the impending 150th anniversary of Neptune's discovery in 1996, archivists at the Royal Greenwich Observatory discovered – much to their surprise – that the 'Neptune file' was missing. These documents include the correspondence between Adams and Airy, and is the only proof of Adams' role in the co-prediction. Upon closer investigation it emerged that the historical material mysteriously disappeared in the mid-1960s, and that there was no indication of it whereabouts. The lost Neptune file was recovered by chance in 1998, during clearance of the office of the late (stellar astronomer) Olin J. Eggen, at the southern hemisphere observatory in Chile. It was found together with several other old and rare books borrowed from the Royal Greenwich Observatory, and in 1999 the documents and books were placed back in the RGO archives. Neptune deserved some more action.

PLANETS ON A DIET

Until now it has been assumed that distances among celestial bodies are so large that their physical size can be safely neglected. As far as gravitation is concerned they are treated as material points, or 'point masses'. All that is required to study their orbital evolution is to place the correct values of the masses in Newton's equations and initiate celestial mechanics. It is a reasonable approximation when dealing with the motion of the planets, which look like bright stars among the constellations, even to the naked eye. It becomes highly unsatisfactory when most of the natural satellite systems orbiting close to their home planet are taken into consideration, while studying the orbital evolution of artificial satellites and in general whenever a close approach between celestial bodies occurs. There are, in fact, major consequences for celestial mechanics related to the shape and finite size of a celestial body.

Although it can be shown that the gravity field surrounding a spherical massive body is the same as if the whole mass were collapsed into its centre, there are only very few examples of this kind in the Solar System. The reason is that, strange it may sound, on an astronomical scale even the Earth cannot be treated as a rigid body but rather following the behaviour of fluids. A rotating drop of water tends to become flattened along its axis of rotation, and this is exactly what is observed for the planets. Jupiter and Saturn are striking examples. The oblateness of a celestial body is measured by the difference between the equatorial radius (the radius of the great circle defining the equator) and the polar radius (the meridian circles passing through the poles). For the Earth this difference amounts to 6,378–6,357 = 21 km. Not much indeed, and it can hardly be seen by simply looking at our planet from space. But an artificial satellite senses even the smallest irregularities in Earth's gravity field, resulting in periodic and secular perturbations of its orbit (Figure 1.15). A new branch of celestial mechanics – space geodesy – represents a modern approach to an ancient problem: the determination of the figure of the Earth by analysing the perturbations acting on the orbits of artificial satellites.

As in the case of Neptune, it is a matter of 'reversing' perturbations, and the ability to build dedicated satellites has resulted in a highly successful collaboration between the US and Italian space agencies. The basic scenario is to place in orbit very simple and symmetric satellites designed to minimise the effect of non-gravitational perturbations such as atmospheric drag and solar radiation pressure. High-precision tracking from the ground can then detect the tiniest perturbations, leading to highly accurate gravitational modelling of the Earth. The satellite design which satisfied these requirements at best turned out to be a sphere only 60 cm in diameter, covered with retroreflectors – high-efficiency optics able to concentrate and send back to Earth the light of a laser beam. The LAGEOS (LAser GEOdetic Satellite) resembles the mirror balls used in discotheques to create light effects. So far two of them have been launched (in 1976 and 1992), and work nominally. Their position can be determined to an accuracy of a few centimetres.

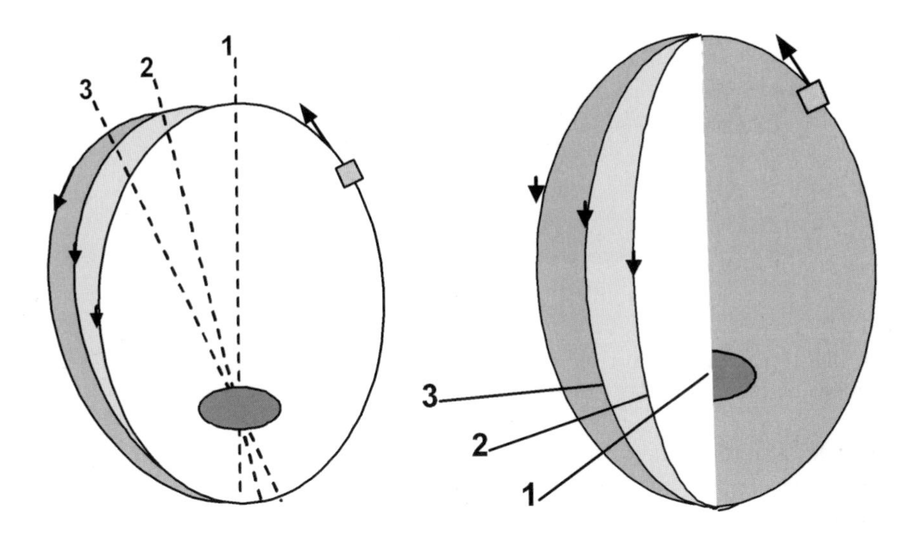

FIGURE 1.15. The motion of a spacecraft around an oblate planet can be visualised as the sum of two different effects: a rotation of the orbit in its own plane (advancement of the pericentre, left plot) and a rotation of the orbital plane in space (precession of the nodes, right plot).

Analysing the complex interrelation between shape and mass distribution has resulted in a highly accurate model of the Earth: the 'geoid'. The conclusion is that not only is our planet definitely non-spherical, but its shape can be only approximated by a regular geometrical figure such as a flattened ellipsoid. Once imaged in three-dimensional high definition the Earth's geoid exhibits 'bumps' and 'holes' corresponding to local or regional mass concentrations and depletions.

GONE WITH THE TIDES

The Moon is the only natural satellite of the Earth, and it is common experience that it does not appear as a bright dot in the sky. The same applies when looking at the Earth from the surface of our satellite, as witnessed by the superb pictures taken by the Apollo astronauts. Earth and Moon 'see' each other as bodies of finite dimensions, also through the eyes of gravitation, thus accounting for one of its most fascinating consequences: explaining the origin of tides.

The existence of a connection between the oceanic tides and the motion of the Moon has always been familiar to mankind; yet the exact mechanism was not known until Newton's enlightening vision of a Universe of attracting masses. Gravitation can be applied to a body of finite size by breaking its mass into pieces and by computing the contribution from each of them. In the case of the Earth, the strength of the lunar attraction depends on the actual distance of the Moon

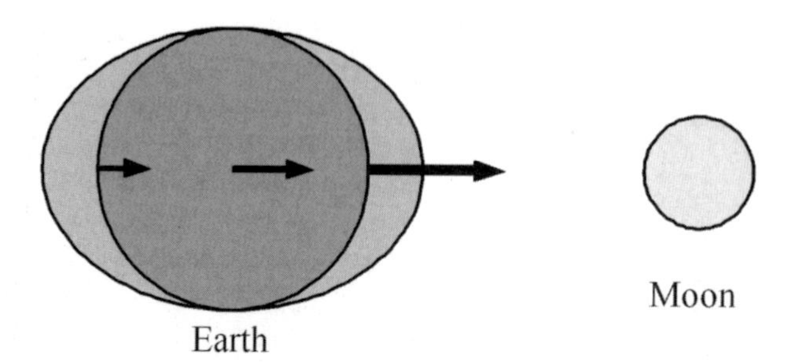

FIGURE 1.16. Tides are the result of the different distance from the Moon of different regions on the Earth, which are therefore subjected to different attractions.

from the different parts in which our planet can be ideally divided. As an example: the region on the Earth's surface immediately below the Moon is more attracted than the centre of our planet, because it is on average 6,378 km (Earth's mean radius) closer to our satellite (Figure 1.16). An inverse reasoning applies to the antipodal region, one Earth radius farther from the Moon.

As the Earth rotates, any region of its surface undergoes a periodic variation of its distance, and then of its attraction, from the Moon. The oceans – being more free to rearrange their mass distribution than the solid ground – form permanent bulges. As seen from the surface of the Earth this results in a periodic raising and lowering of the sea level observed twice per day. The observation that the maxima and the minima of the tides do not occur in close alignment with the position of the Moon in the sky is only a consequence of the tidal bulges being carried along by the rotation of the Earth.

The Earth–Moon system is a neat example of tidal interaction, because the observable effect is amplified by the presence of liquid masses on our planet. But tides are much more complex. Rocky surfaces also experience strains and stresses known as 'solid tides', and the consequent deformations are of far lesser amplitude, amounting, in the case of our planet, to only a few centimetres. Precise measurements of the figure of the Moon show that our satellite presents a tidal bulge in the direction of the Earth generated by the gravitational pull of our planet. Solid tides – even if less 'spectacular' – are present almost everywhere in the satellite systems of the giant planets, providing the basic mechanism for explaining a variety of different phenomena observed on their surfaces, including highly energetic events. The break-up of small bodies during close encounters with the planets is often due to the tidal stress exceeding internal cohesion forces. This was clearly demonstrated by comet Shoemaker–Levy 9, which was torn apart following a close approach to Jupiter in July 1992, the string of fragments (Figure 1.17) finally plunging into the planet's atmosphere two years later.

Finally, the tides are also responsible for the long-term orbital evolution of a system, and are therefore one of the most important perturbations in celestial mechanics, as they slowly modify the rotation and the revolution periods of the

FIGURE 1.17. The disintegration of comet Shoemaker–Levy 9 was produced by the tidal stress due to an extremely close approach to Jupiter. (Courtesy NASA/HST.)

interacting bodies. The friction caused by the tidal bulge, whether solid or liquid, tends to oppose the rotation of a body, thus slowing it down. The duration of the 'day' of a celestial body undergoing tidal interaction therefore slowly increases. On the other hand the existence of a tidal bulge modifies the symmetry of the mass distribution. The resulting perturbation causes two tidally interacting bodies to drift away from each other as a consequence of the steady increase of the size of their orbits. As an example: the duration of the terrestrial day at present increases by approximately 0.001 of a second per century, while the lunar month (its period of revolution around the Earth) slowly lengthens. These effects – observed for the first time in 1693 by Edmond Halley, on the basis of ancient eclipse records – have minor consequences for human everyday life, but in the greater scheme they become dominant.

The process of slowing rotation and enlarging orbits will continue until the Earth's period of rotation equals the Moon's period of revolution. It has been computed that this will happen some time in the distant future when the length of both the day and the lunar month have reached a common value of about 40 present Earth-days. The Earth will then permanently show the same face to the Moon, with a tidal bulge firmly oriented toward our satellite. Probably no further friction will occur, and the system will be forever locked by tides.

2

Three bodies and no solution

> Science is always wrong. It never solves a problem without creating ten more.
>
> *Bernard Shaw*

One of the most intriguing science fiction novels written by Isaac Asimov – *Nightfall* – includes much concerning celestial mechanics. The action takes place on a planet orbiting inside a multiple star system. Because of the rare occurrence of the night on the planet, the habits of the people are quite peculiar – especially as far as psychological and social aspects are concerned. Science is also deeply affected. Due to the joint action of several massive stars the orbit of the planet is far from resembling an ellipse, and gravitation becomes difficult to discover, even for an alien Newton. So how can be explained, in a simple way, the entangled trajectories of a planet driven by the simultaneous attraction of several stars? – especially knowing that an incredibly high level of complexity is reached, even if we restrict our attempts to the orbital paths resulting from the mutual influences of only three bodies.

CELESTIAL MECHANICS GET THE BLUES

The aim of the physical science is to observe nature at work and to describe its behaviour. The language of physics is mathematics, because it allows the development of abstract models mimicking reality: the *dynamical systems*. For example, a pendulum is a simple dynamical system with motion in the form of oscillations. Conversely, it is not easy to imagine the dynamics of a neural network such as that acting in the human brain, because its 'motion' is represented by the variation of the status of each neuron. A neural network is a dynamical system far more complex than a pendulum.

The description of dynamical systems relies on mathematical equations in which their basic characteristics are coded and unknown key quantities are identified. Finding the solution to these equations corresponds to unveiling the motion of the system, which in turn requires the prediction of its future behaviour and the determination of its past history. The process which allows us to find the desired solution is called *integration*. A dynamical system that admits a mathematical solution is said to be *integrable*, otherwise we classify the system as *non-integrable*.

Unfortunately the determination of the laws of motion requires sophisticated mathematical techniques, and in many cases a general solution might not even exist. This is the reason why before starting to search for a solution we should try to determine whether or not the system is integrable. But again this is not an easy task; nor it is always possible.

The two-body problem is an integrable system and therefore has explicit solutions to the equations of motion which correspond to three types of orbit: ellipse, parabola and hyperbola. Relying on this encouraging result it might be thought that when adding just one more celestial bodies (for example, the Sun–Earth–Jupiter system) the resulting motion should not be too difficult to integrate. But the statement of the problem is misleadingly simple, as witnessed by the frustrated efforts of many skilled scientists who have tried to find a general solution to the three-body problem.

In this respect there is an enlightening story concerning the prize awarded by King Oscar II of Sweden and Norway to one of the greatest celestial mechanic of all times: the French mathematician Henri Poincaré (1854–1912) (Figure 2.1). It began in 1886 when the Swedish mathematician Gosta Mittag-Leffler suggested to King Oscar – who wanted a spectacular event to celebrate his 60th birthday – that he offer a prize to whoever could answer some of the highest scientific questions of the time. In this way the name of the King would be forever associated with the advancement of science. Among the selected items there was the solution of the 'N-body problem' (where N indicates that there are more than two gravitationally interacting bodies).

Henri Poincaré – who at that time was already a distinguished scientist – was encouraged to apply by Mittag-Leffler himself, and he immediately began to work on the subject. He tackled the three-body problem first, and quickly developed some innovative methods and tools to deal with the complexity of the problem. As an example, the mapping technique and the importance of periodic

FIGURE 2.1. Henri Poincaré at different stages of his life.

orbits (which will be treated in the following sections) were introduced. Although Poincaré produced many insights into the nature of the three- and N-body problem, the deadline for the prize - 1 June 1888 - was approaching, and he still did not have any general statement on the subject. Eventually he succeeded in producing a stability result valid for a three-body system and liable to be extended to the N-body case. This was a remarkable conclusion, drawn after the almost 160 pages of high-level celestial mechanics. On 21 January 1889 Poincaré won the prize for these great advances - although he did not solve the N-body problem.

Poincaré then began to revise his book-sized article, which was to appear in a special issue of the scientific journal *Acta Mathematica*. It took a considerable time to edit - almost a year. Then, on 30 November 1889, when everything seemed satisfactory, Poincaré sent a telegram to stop the printing of the article because there was a mistake. He realised that he had been wrong in his general stability result! This was a rather embarrassing situation, and he took financial responsibility - but the amount it cost him was more than the money he had received with King Oscar's prize! Yet for such a man of science, reputation was worth more than money. Poincaré renewed his efforts in investigating the three-body problem, and after some months of frantic hard work he submitted a new version of his prize-winning paper, which by then had grown to 250 pages. In it he dismissed any statement of general nature on the solution of the three- and N-body problems and introduced a revolutionary concept in celestial mechanics: chaos.

The Mittage-Leffler files

In the year between the first and the second printing of Poincaré's memoir on the three-body problem, the situation became almost paradoxical. Even if the revised article was even more important for celestial mechanics, the strongest conclusion which allowed him to win the prize was no longer valid. Should the King ask Poincaré to return the award? No official steps were taken. Possibly in an attempt to hide what was happening, Gosta Mittag-Leffer, as editor-in-chief of *Acta Mathematica*, searched for the copies already distributed and asked for them back with the promise that he would replace them with the new edition. Lacking documentation in the years to come, many historians of science wondered about the details of Poincaré's famous mistake.

Almost a century later, in 1985, Richard McGehee, of the University of Minnesota, spent a sabbatical year in Djursholm, near Stockholm. There, the mathematical research institute is set in a beautiful villa – the former residence of Mittag-Leffler, which he had donated together with his precious archives. As usually happens only in the movies, McGehee found an uncatalogued dusty file containing the only surviving copies of the first printing of *Acta Mathematica* in which Poincaré's original prize-winning paper was published. On one of them is handwritten, in Swedish: 'The whole edition was destroyed – M.L.'.

The core of Poincaré's thought is that when three or more bodies are gravitationally interacting a wide variety of orbital regimes becomes available. Open trajectories, close encounters, collisions, resonances and all intermediate and transition patterns from one case to another: chaos. A general solution which encompasses them all would require an astonishing complex formulation. Poincaré's great contribution was the introduction of qualitative methods, which may not provide an explicit solution but, as we shall see, nevertheless allow us to gain a deep understanding of the behaviour of the dynamical system under study.

From a strictly mathematical point of view it can be said only that the difference between the two-body and the three-body problem is that the border between integrable and non-integrable systems has been crossed, and that there is no way back.

LAGRANGE IN EQUILIBRIUM

The non-existence of a *general solution* to the three- and N-body problem does not prevent the motion of a system being accurately described in some specific cases: the *special solutions* that we are about to describe. The Italian-born mathematician Joseph-Louis Lagrange (1736–1813) was a pioneer in this field. In 1772 he proved that three bodies of arbitrary mass can keep their relative configuration unchanged while moving along their orbital paths. There are only two geometries which satisfy the Lagrange conditions: either the three bodies lie along the same line, or they are located at the vertices of an equilateral triangle (Figure 2.2). The positions in space of the third body satisfying these special solutions are called *Lagrangian equilibrium points.* There are five of them: the *collinear* L_1, L_2, L_3 and the *triangular* L_4 and L_5.

The existence of these peculiar configurations can be understood by recalling the familiar geometries of the two-body problem. Let us suppose that a body of mass m orbits around another body of mass M, and that we are inside a small

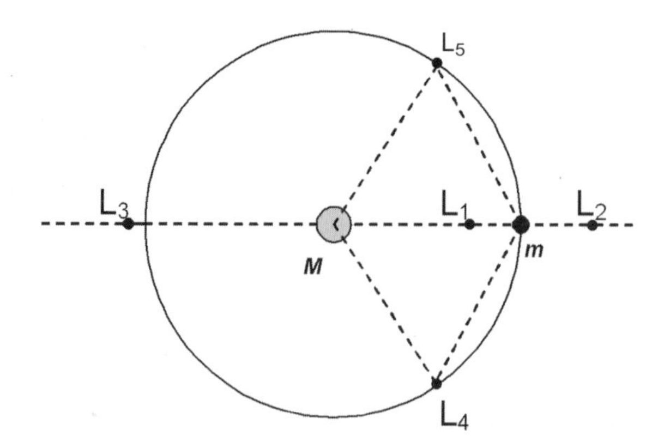

FIGURE 2.2. Location of the Lagrangian equilibrium points.

spacecraft acting as the third body of the system. If our position happens to be somewhere along the line joining *M* and *m*, the spacecraft 'feels' the gravitational attraction exerted by both the massive bodies in opposite directions. It is then possible to split the three-body system *M*–*m*–spacecraft into a 'double two-body problem'. The first is represented by the pair *m*–*M*, and the second is obtained as follows. The spacecraft can be thought of as moving around a body with a mass somewhat less than *M*, because its gravitational attraction on the spacecraft is weakened by the presence of *m*, which pulls in the opposite direction.

In general this simplification is valid only instantaneously, since the alignment breaks up as the bodies move along their orbits. Yet by applying Kepler's third law to each of the two-body subsystems, the corresponding periods of revolution for all values of the mean distance between them can be computed. It is then possible to look for a location along the *M*–*m* line such that in the 'double two-body' approximation the periods of revolution of both *m* around *M* and the spacecraft around the 'reduced' *M* are the same. This would make *m* and the spacecraft revolve at the same angular speed, thus preserving the alignment forever. In other words, it is an attempt to synchronise the two orbital motions taking into account that the spacecraft feels a smaller mass in *M*. The solution of the problem was determined by Lagrange, who provided the mathematical equations needed to compute the place of the spacecraft: at the equilibrium point L_1. Similar considerations also hold for the other collinear points, with the notable difference that in L_2 and L_3 the spacecraft feels a larger mass in *M* because now *m* pulls in the same direction.

In describing the dynamics of the triangular equilibrium points L_4 and L_5 it is again useful to separate the system into a double two-body problem. In this representation the spacecraft shares the same orbit of *m* around *M*, just displaced 60° ahead (L_4) or behind it (L_5) (Figure 2.2).

The computation of the position of the Lagrangian points involves only the masses of the celestial bodies, and since our Solar System can be split into several three-body subsystems (for example, planet–satellite–Sun, Sun–planet–asteroid) the regions close to the Lagrangian points are always worth investigating. The discovery of asteroids following and preceding Jupiter on its orbit around the Sun, while keeping their distance from the planet as predicted by Lagrange, was the first confirmation of the practical importance of the triangular equilibrium points. Since then, many other Lagrangian-like configurations have been observed. In Saturn's system some of the large natural satellites are accompanied by tiny moonlets either in L_4 or in L_5, while many small satellites dance with the rings, stepping from one Lagrangian point to another. The possibility of moving back and forth from L_4 to L_5 and passing through L_3 identifies a new type of orbit called 'horseshoe', because of the peculiar shape when drawing it in a rotating reference frame (see, for example, Figure 2.3). Horseshoe orbits are sufficiently stable to host small celestial bodies for a relatively long time. This is the case for the asteroid Cruithne – the Earth's secret companion, which has been discovered sharing almost the same orbit of our planet (Figure 2.4).

As far as spaceflight dynamics is concerned, the Lagrangian points have been

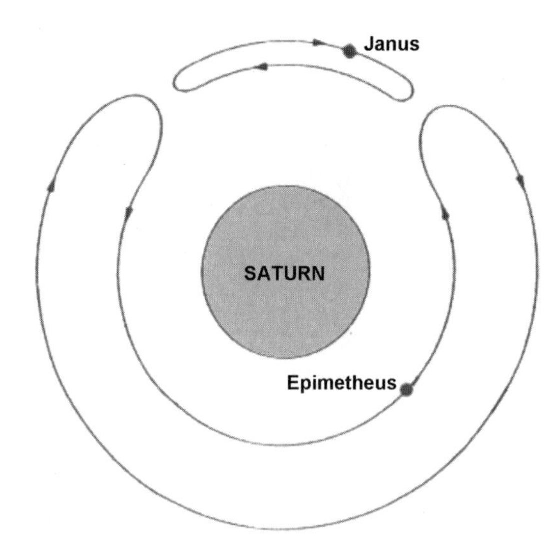

FIGURE 2.3. The dynamical mechanism which prevents the two co-orbital satellites Janus and Epimetheus from collision results in horseshoe-like orbital patterns, which clearly appear when their motion is plotted in a Saturn-centred reference frame rotating with the mean angular velocity of the two bodies.

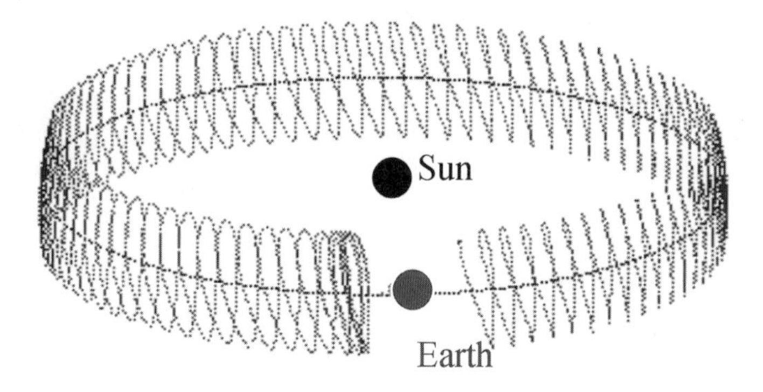

FIGURE 2.4. The high-amplitude horseshoe orbit of asteroid Cruithne.

widely used for 'parking' man-made celestial bodies in space. As an example, the European SOHO (SOlar Heliospheric Observatory) has been stationed for many years around the Sun–Earth L_1. It observes our star 24 hours per day, 365 days per year, without the interruptions that would necessarily occur while flying on the night side of our planet.

The first European scientist

Giuseppe Luigi (Joseph-Louis) Lagrange was born in Turin in 1736. There he studied at university, and began to teach mathematics at the local military school. In 1766 he moved to Berlin where, succeeding Euler, he was appointed Director of Mathematics of the Academy of Sciences. He worked in Berlin for 21 years – a period encompassing almost all of his major discoveries. At the age of 52 he became a member of the Bureau des Longitudes (Paris), and in 1797 he was appointed professor at the Ecole Politechnique. He remained in Paris until his death in 1813, and is buried in the Panthéon. His life somehow represents the perfect circle of a truly European scientist.

PERIODIC ORBITS

Among the special solutions of the three-body problem, particular attention must be paid to the existence of *closed trajectories* – orbital patterns that repeat themselves at regular intervals of time. Orbits of this kind have been found, and bear the generic name of *periodic orbits*.

The problem here is to determine the values of the position and of the velocity of a celestial body, such that after a certain period of time an exact repetition of its orbital pattern occurs – no matter how complex. A useful strategy consists in searching for symmetrical configurations. An ellipse – a known periodic solution of the two-body problem – is a straightforward example of a symmetric periodic orbit. Imagine drawing only half of the ellipse from pericentre to apocentre, and placing a mirror along the semimajor axis. The complete ellipse is restored. Similarily we can deal with more complex trajectories by trying to understand where to 'place the mirror' in order to cause the reflections that produce a periodic orbit. This goal is, of course, pursued using mathematical techniques, and in particular by applying the *mirror theorem* developed by the British astronomer Archie E. Roy and the Canadian astronomer Michael W. Ovenden in the early 1950s. Following in Poincaré's footsteps they tried a qualitative approach by determining the geometrical conditions in the motion of the celestial bodies which represent 'virtual mirrors', in the sense that the future evolution of the system is a mirror image of its past. These 'mirror configurations' do indeed exist, and are characterised by a peculiar condition: all the velocities of the bodies must be perpendicular to all the lines that can be drawn from one body to the other. Upon closer investigation it can be shown that there are only two possible geometries satisfying a mirror condition. The first occurs when all the bodies are aligned with all velocities perpendicular to their common line (Figure 2.5); the other when all the bodies lie on the same plane, with all velocities perpendicular to that plane.

These are not abstract geometries with no connection with reality. As an example, the collinear case causes the occurrence of a well-known astronomical event: an eclipse (Figure 2.5). In general the mirror theorem implies the

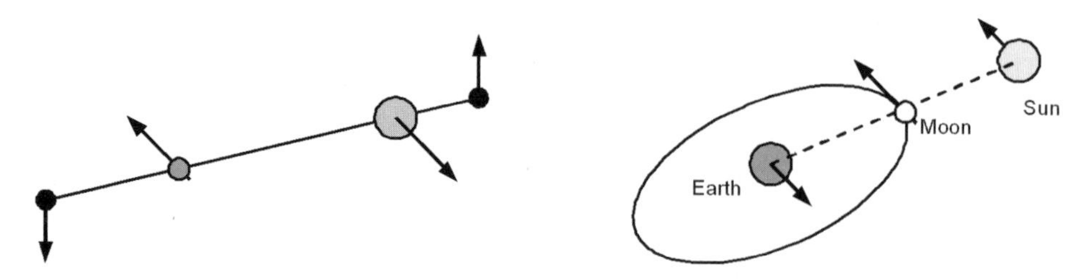

FIGURE 2.5. A collinear mirror configuration (left) and solar eclipse geometry (right).

statement that when at a given time N celestial bodies pass through a mirror configuration (no matter how crazy their trajectories) their motion is symmetrically reversed. This is certainly useful in studying non-integrable systems, as it provides a way of knowing the past or future evolution of an N-body system without the heavy computational load which is usually involved in determining its dynamical evolution.

Yet the real breakthrough is one step ahead when two subsequent mirror configurations are isolated. Two reflections cancel each other, so that for celestial mechanics the system returns to the beginning. In other words, we have found a periodic orbit with a period of twice the time spent between the occurrence of the two mirror configurations.

MAPPING CHAOS

The dynamical behaviour of an integrable system, such as the two-body problem, can be predicted with high accuracy once the mathematical solution of its equations of motion is determined. Conversely, chaotic behaviour is typical of non-integrable systems and becomes increasingly dominant as the complexity of the system grows.

The three-body problem is not integrable, and thus we can expect that sooner or later chaos will arise. Let us consider the dynamical system formed by the three most important objects of our planetary system: our star, the Sun, our home planet the Earth, and the giant massive Jupiter. Jupiter's mass is about 300 times that of the Earth, and its perturbative effect on the orbital motion of our planet cannot be neglected. In order to study the magnitude and the consequences of this perturbation we introduce a step-by-step procedure. Suppose, as a first approximation, that Jupiter suddenly disappears from our model. The system is then back to the well-known interaction between the Earth and the Sun – a two-body problem described by Kepler's laws. A fake Jupiter having a negligible mass, like that of a cricket ball, is then introduced. Obviously it has no detectable influence on the motion of the Earth; but if we increase the mass of the pseudo-Jupiter, still keeping it small enough, the motion of the Earth will slightly change with respect to the integrable case. The trajectory is no longer a closed ellipse, still resembling it at first glance. This is an example of a

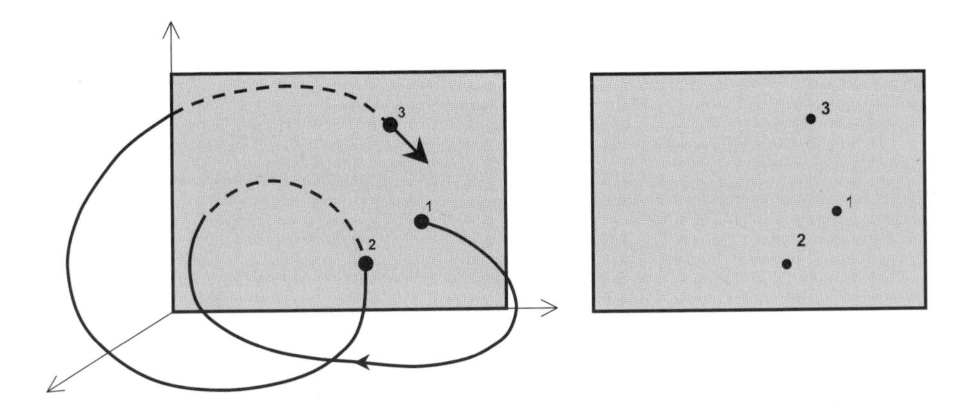

FIGURE 2.6. How to build a Poincaré map. The intersections of a trajectory with a given plane are recorded, and their arrangement provides information on the dynamics of the system.

quasi-integrable system, as it still forms the basis of the two-body motion. An approximate mathematical solution can in general be found using *series expansions* – long mathematical expressions containing an infinite number of terms which can be truncated when the desired accuracy is reached. The mass of the pseudo-Jupiter, the value of which can ideally be changed, is called the *perturbative parameter* to indicate that the corresponding trajectories of the system are obtained as perturbations of a Keplerian ellipse.

What happens when the mass of the third body is allowed to grow without restriction? The overall scenario becomes more and more complicated, involving many different types of motion from regular to chaotic. The revealing of the workings of this labyrinth is highly desired. Indeed, the development of a general approach to describe the behaviour of a system for any value of the perturbative parameter is one of the major contributions of Poincaré's work. It bears some analogy with a geographical map in the sense that it displays the location of different dynamical regions, whether regular or chaotic. Mapping is an essential tool to avoid becoming lost among the many orbital regimes accessible to the system.

We shall show how it works with a practical example based on three bodies. Suppose we analyse the motion of the Earth under the influence of the Sun and Jupiter. Imagine 'cutting' a given trajectory with a plane having a preassigned orientation in space (Figure 2.6) and marking the location of the intersection points. Keeping the plane fixed and repeating the same procedure for different values of initial conditions, characteristic patterns will soon appear (Figure 2.7). The plots obtained in this way are called *Poincarè maps*, and their intepretation is relatively simple: When the intersection points trace regular curves an ordered motion is involved, while regions irregularly filled by dots are representative of the onset of chaotic regimes. It must be stressed that regular motions include

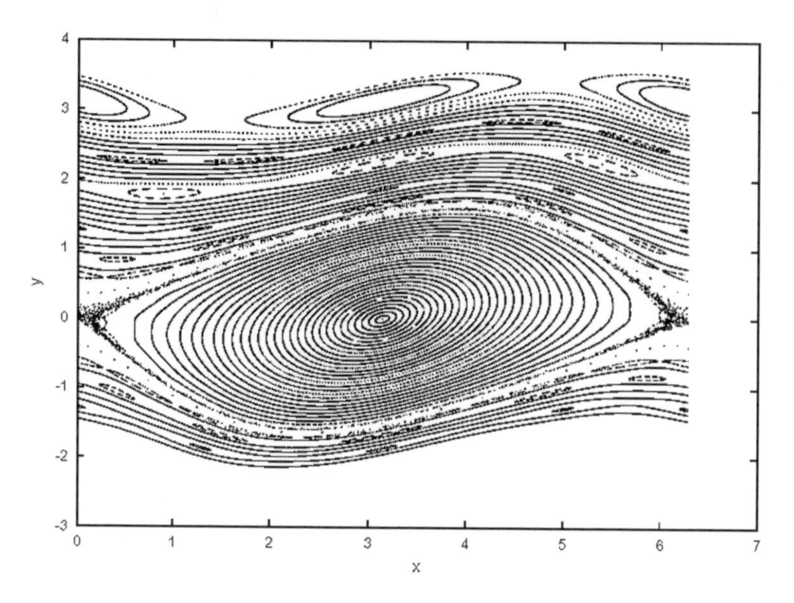

FIGURE 2.7. The standard map that describes the dynamics of a pendulum subjected to a periodic perturbation.

A chaos machine and the butterfly effect

In order to grasp the exact meaning of 'chaotic motion' it is possible to show how to build a simple machine exhibiting an extremely complex behaviour: a *chaos machine.* Take a bicycle wheel and fix it to the ground so that it can rotate freely, and on each spoke hang a small plastic container with a hole at the bottom. Then pour a steady flux of water over the wheel so that it fills the containers as they pass under it. The different weight of the containers, filled in turn by water and emptied through the holes, drives the rotation of the wheel. What will be the resulting motion? Will the wheel always turn in the same direction at more or less constant speed, or not? Will it remain still or oscillate back and forth? That is, will its motion be ordered or chaotic?

A couple of considerations are meaningful: the higher the speed of the wheel, the shorter the time spent by each container under the water flow, thus accordingly reducing the refill. The rate at which the water is lost through the holes is also dependent on the rotation and whether the container is climbing up, going down, or tangential to gravity. The combination of these opposing (but totally deterministic) effects causes the wheel to rotate at different velocities and to stop and reverse its motion in an apparently random sequence. Even worse, it is impossible to reproduce the same behaviour, even if we carefully attempt to restart the wheel with the same initial conditions. Every time we try, it rotates as if at its own will. This happens because of the extreme sensitivity of the system as a whole. The slightest variation in the water flow or in the initial position of the spokes soon leads to huge variations in the rotation of the wheel.

both periodic orbits as well as quasi-periodic trajectories, which are solutions of the equations of motion which never exactly retrace themselves, though approaching indefinitely close to the initial conditions at regular intervals of time.

In conclusion, the powerful method introduced by Poincaré exploits the trace of the trajectories on a plane to describe the general dynamical properties of a system. The efficiency of the mapping technique has been greatly improved by the use of modern computers, and it is now widely implemented to describe the long-term orbital evolution of celestial bodies.

In 1963 the meteorologist Edward Lorenz introduced a concept usually referred to as the *butterfly effect*, by which extremely small perturbations – such as those triggered by the flap of a butterfly wing – can cause huge weather changes on a global scale; a high sensitivity to small displacements of the initial conditions. Chaos hides in the consequent difficulty of predicting the motion of the system.

KAM AND ALL THAT

We are now able to recognise chaotic orbits from the way in which they intersect a Poincaré map. The next questions are: how does chaos relate to stability?... and does a chaotic regime necessarily lead to instability? Whenever the orbital motions are regular, stability is typical, while a certain degree of uncertainty is introduced when chaos occurs. This does not necessarily imply catastrophic events, like collisions among the celestial bodies involved or their ejections, but rather a much higher sensitivity of the trajectories to small perturbations. As a consequence this is a strong impact on the mathematical and numerical modelling of the dynamical evolution of the system – which always relies on some approximations – demanding a higher accuracy. In order to investigate the stability it is therefore essential to have an exhaustive description of all possible future evolutions of the system, including those characterised by chaos.

In order to render this concept more concrete, let us consider an astronomical application: the motion of an asteroid subject to the gravitational attraction of the Sun and to the influence of a third body that we again call a 'pseudo-Jupiter'. The perturbing parameter is represented by the ratio of the masses of the pseudo-Jupiter and of the Sun. As long as the mass of the pseudo-Jupiter remains sufficiently small, the trajectory of the asteroid is regular, being very close to a Keplerian two-body ellipse. As far as the mass of the perturber raises, the motion of the asteroid becomes increasingly perturbed and loses regularity whenever the perturbing parameter reaches a critical value, which marks the transition to chaos. Computation of this value is therefore the first step in determining stability.

Numerical and rigorous mathematical methods have been developed to compute the critical value of the perturbing parameter. The KAM theory belongs to the latter class, and is one of the most commonly used. The acronym derives from the names of three mathematicians who contributed to its development towards the middle of the twentieth century: Andrei N. Kolmogorov (1903–1987), Vladimir I. Arnold (1937–) and Jurgen Moser (1928–1999) (Figure 2.8).

KAM theory provides a tool to investigate the stability of a quasi-integrable system when traditional perturbation theory fails. In some cases (for example, resonant trajectories or chaotic motions) the occurrences of what mathematicians call 'diverging series', 'small denominators' and 'singularities' prevent us from obtaining reliable results. We therefore need to develop new ideas to investigate the dynamics of quasi-integrable systems and, in particular, to study the stability of the three-body problem. The breakthrough came during the International Meeting of Mathematics, held in Amsterdam in 1954, during which Kolmogorov presented a new result, which over the ensuing years was extended by Arnold and Moser.

Without entering into the details, it can be said that KAM theory has the advantage of overcoming the difficulties arising from classical perturbation theories, thus providing proof (under very general mathematical assumptions) of the stability of the motion of a system for an infinite time. The disadvantage is that in order to apply KAM theory the perturbing parameter must often be unrealistically small. As an example, recalling the three-body system discussed above, a KAM-guaranteed stability requires that the mass of the pseudo-Jupiter should not grow over a limiting value much smaller than the 'real' Jupiter. This means that the motions are always ordered and stable for values of the perturbing parameter (the mass ratio of Jupiter and the Sun) less than a given value. KAM theory simply provides a lower estimate of the chaotic transition. The farther we go from a safe KAM estimate, the easier it is to become caught in chaotic motion.

FIGURE 2.8. A.N. Kolmogorov, V.I. Arnold and J. Moser.

In this respect KAM theory can be compared to the trials for testing the road-holding power of a vehicle. The general stability of a car travelling on a winding road is granted only for relatively low velocities. Once a certain value is achieved, stability depends upon the skill of the driver; but it eventually becomes impossible to keep control of the car, even by a professional race driver. KAM is a sort of high-speed driving school for celestial mechanics.

To be more exhaustive, it should be mentioned that the applications of the original versions of KAM theory were able to prove the stability of the three-body problem provided the mass ratio is less than 10^{-48} – equivalent to replacing Jupiter with something with a mass billions of billions times smaller than a tennis ball! It is not surprising to find that such a system is stable, since the pseudo-Jupiter is essentially a ghost and its influence on the asteroid is definitely negligible. However, it must be stressed that the original versions of KAM theory were oriented to obtain a theoretical result rather than to practical implementation. Nevertheless, from the astronomical point of view, although satisfied by its beautiful mathematics, we prefer a KAM theory that provides results consistent with astronomical measurements.

A new approach to this problem followed the progress of computers in recent decades. Scientists were intrigued to find a synergy between mathematical theories and computer programs. A new technique proving mathematical theorems with the aid of a computer has been developed in recent years, and has been widely used in several fields of mathematics. As we know, computers work with limited precision, since all quantities are represented by a finite number of decimal digits. Therefore, any result produced by the computer is affected by the *rounding* errors, which eventually spread, to be replaced by *propagation* errors. An example of rounding error is the following. If computer precision amounts to three decimal digits, the result of the division of 1 by 7 is equal to 0.143 with an error on the last decimal digit, since the true result is equal to 0.142857142857... (determine the remaining digits yourself!). Working with finite precision, the computer rounds the result up or down. By performing further operations similar to the previous one it is impossible to avoid propagation errors.

Nevertheless, it is possible to keep track of rounding and propagation errors through a technique called *interval arithmetic,* which replaces the finite precision result with an interval which certainly contains the true result. In the previous example, the output is given by the interval (0.142, 0.143), and subsequent operations are performed on intervals.

Finally, it happens that the combination of theory and computers is very effective. The machine allows us to perform a huge number of computations, and the errors are controlled through interval arithmetic, which retains the validity of the mathematical proof. The new strategy obtained combining KAM and interval arithmetic allowed one of the authors, in collaboration with Luigi Chierchia (University Roma Tre), to obtain results in agreement with the physical measurements, thus bridging the gap between the rigour of the mathematical computations with the certainty of the astronomical observations.

INSIDE THE RINGS OF SATURN

James Clerk Maxwell (1831–1879) owes his celebrity to his discovery of a common origin in electrical and magnetic forces. In doing so he discovered electromagnetic waves, which now rule many aspects of our life, from radio and TV broadcasting to microwave radiation, and from X-rays to laser beams. But the first recognition of his genius came from celestial mechanics.

As a young scientist, in 1856 Maxwell won a prize awarded by the University of Cambridge with the aim of unveiling the true nature of Saturn's rings (Figure 2.9). At that time it was a longstanding problem. More than two centuries earlier Galileo had been the first to recognise that there was something strange about Saturn. Because of the limited resolving power of his telescope he could only say that the planet seemed to be accompanied by two large satellites, one at each side, that surprisingly did not move around the planet. Eventually they had mysteriously disappeared. Some time later, in the second half of the seventeenth century, the Dutch astronomer Christiaan Huygens (1629–1695) built much better telescopes, which allowed him to affirm, without any doubt, that Saturn was surrounded by a large ring. Further progress was made by the Italian-born founder of the Paris Observatory, Gian Domenico Cassini (1625–1712), who detected a division in the ring, which now bears his name.

The increasingly detailed observations called for a physical explanation. Maxwell showed that the ring could not consist of a single layer, whether solid or liquid, because even a small perturbation would immediately break it into pieces. Saturn's rings were instead the result of a multitude of small celestial bodies orbiting the planet. In Maxwell's view, every ring is like a pearl necklace encircling Saturn, and each pearl is a small satellite in orbit around the planet.

Today we know that the dimension of the celestial bodies in Saturn's rings varies from kilometre-size boulders to dust grains. Their motion is not Keplerian because of the perturbations exerted by the large natural satellites of the planet, which ultimately determine the large scale structure of the rings. This is why the rings of Saturn can be considered a full-scale laboratory for celestial mechanics.

FIGURE 2.9. Galileo's drawing, the young J.C. Maxwell, and the fine structure of Saturn's rings as imaged by the Voyager 2 spacecraft in 1981. The two main ring systems are separated by the large Cassini division, recognisable in the Voyager image as a dark belt. (Courtesy NASA/JPL.)

The evolution of thousands of three-body (planet–satellite–ring particle) and N-body systems can be studied in detail and the theories of motion checked with observations.

Many of the prominent features of the ring system – such as the divisions (regions apparently devoid of particles) – have been successfully explained by celestial mechanics in terms of the dynamics generated by one of the following three-body problems: a ring particle under the effect of Saturn and of a satellite, or a ring particle driven by the gravitational attraction of two small satellites. The detailed images sent by the Voyagers and by the Cassini orbiter provide an incredibly new amount of open problems. Relying upon them, a new class of celestial bodies has been born: the moonlets. These tiny irregularly shaped objects orbiting among the rings play an essential role in shaping the structure of the system. There are many examples of this kind.

The F ring is a very narrow ring located just outside the main system. The small particles which it comprises are confined by the presence of two orbiting moonlets – one on each side of the ring. Prometheus and Pandora are referred to as 'shepherd' satellites, because without their attraction the particles of the F ring would soon escape due to the perturbations of the large satellites of Saturn.

A 10-km moonlet orbiting inside Encke's division was discovered due to the prediction of celestial mechanics that told astronomers where to look for it. It appeared as a barely distinguishable faint dot on some Voyager images, and therefore remained unnoticed for almost 10 years. It has been given the name of the Greek god Pan.

Janus and Epimetheus are known as the *coorbital satellites* because they travel along almost identical orbits. Following Kepler's second law, the one slightly closer to the planet also moves slightly faster, thus slowly reaching its companion. A collision is avoided because during the final approach the mutual gravitational attraction causes the moonlets to exchange their orbits. The predator becomes the prey, and they both slowly drift apart to commence a new four-year cycle, at the turn of their celestial waltz.

A FLASH IN THE NIGHT

Henri Poincarè is considered one of the last 'universal scientists' due to his seminal contributions in many different fields of science. As we have seen, he was a pioneer of celestial mechanics, but this did not prevent him from also being an excellent science educator and a researcher on the psychology of scientific creation. In trying to understand where great ideas originate, he wrote the aphorism: 'A thought is a flash between two long nights, but that flash is everything.' By discussing in detail one of his best 'flashes' – known as Poincarè's conjecture – we close this chapter dedicated to the three-body problem – the very first and almost unsolvable difficulty that is met when the Keplerian approximation of two bodies moving in elliptic orbits is no longer valid.

We left Poincarè busily working out his error and discovering chaos. The huge article that he submitted to *Acta Mathematica* was just the beginning of a lifetime's work that resulted in the publication of his celebrated treatise *Les Méthodes Nouvelles de la Méchanique Céleste* (*The New Methods of Celestial Mechanics*). In Chapter 3, section 36, page 82 there is a statement which can be considered as one of the most enlightening visions of celestial mechanics. It can be confidently said that those few lines spurred much of twentieth-century research on the dynamics of N-body systems. Its influence lasts until today, encompassing the study of the trajectories needed for the automatic and human exploration of the Solar System. For simplicity we present Poincarè's vision as three separate statements:

> 'Here is a fact that I have not been able to demonstrate rigorously, but which appears to me nevertheless very likely to be true.'
>
> 'Being given equations of the form defined in [Section] 13 and a particular solution of these equations, one can always find a periodic solution (whose period, it is true, can be very long), such that the difference between the two solutions will be as small as one wishes during a time as long as one pleases.'
>
> 'Besides, that which makes periodic solutions so valuable is that they are, so to speak, the only breach through which we can attempt to penetrate what was previously thought impregnable.'

Initially Poincaré warns the reader that he is taking the risk of saying something that he 'only' believes. But this is not unusual for science. According to the *Dictionary of Mathematical Terms*, a conjecture is 'a statement that, being verified as true in some cases, it was not possible to dismiss as false in any other occasion'. In other words, a conjecture is a stroke of intuition – a deep belief that has sufficient scientific ground to be formulated, but not enough to be demonstrated.

Poincaré then moves on to the core of the matter. For any trajectory, no matter how complex, obeying the laws of gravitation (the 'Section 13' mentioned in the text refers to the N-body equations of motion), it should always be possible to find a periodic orbit as close as we like to an arbitrarily long branch of the original trajectory. The only problem arises from the fact that this periodic orbit could have a very long period.

The final statement was possibly been inspired by the fact that Poincaré lived in a tower in Rue Gay-Lussac, in Paris. It stresses that the extensive use of periodic orbits is the key for investigating the complexity of the three-body problem.

We can now try to translate Poincaré's conjecture in a more user-friendly form:

> 'I am confident that for every *gravitational drawing* traced by celestial bodies, there exists a periodic orbit arbitrarily close to it, but possibly of very long period. Just remember: periodic orbits are useful for solving stiff problems!'

It can naturally be wondered why the brilliant Poincarè could not demonstrate his own statement. The answer is that searching for periodic orbits is not an easy task. When more than two interacting bodies are taken into consideration and there is no solution to the problem, either complex analytical theories or the extensive use of numerical methods are needed. In practice it is often necessary to perform a heavy load of computations, either when developing series expansions or implementing numerical integration techniques. However, even a century after Poincarè's 'flash in the night', a rigorous demonstration of his conjecture is still missing.

Yet modern high-speed digital computers allow extensive exploration of the neighbourhood of a trajectory, hunting for longer and longer periodic orbits

The psychology of mathematical invention

Henri Poincaré's strong personality is illustrated by the many stories concerning his private life. His nephew once said:

> 'Uncle Henri used to think all the time. On the road to the University, while attending some important meeting or during the usual relaxing walks after lunch. He kept on thinking while stepping in the doorway and in the corridors of the Institute with the keys hanging from his hand. He was thinking even sitting at the dinner table with all the family around or reaching the point of leaving in astonishment a friend right in the middle of a conversation only to follow a thought that had suddenly flashed in his mind.'

In his essay on the psychology of invention in mathematics, the French mathematician Jacques Hadamard (1865–1963) discussed the relationship between the conscious and unconscious, as inspired by a seminal paper, delivered at a conference of the French Society of Psychology, by Henri Poincaré – 'the man whose impulse is traceable in all contemporary mathematics'. (Jacques Hadamard, *Essai sur la psychologie de l'invention dans le domaine mathématique*, Paris, 1975.)

Poincaré recorded the circumstances which led him to one of his most brilliant ideas: the 'theory of Fuchsian groups'. After two weeks of determined but unsuccessful attempts, he eventually took a relaxing break to take part in a geological excursion. When he arrived at his destination he decided to take an omnibus, and at the very same moment he stood on the platform there came the idea, although he was not thinking of his mathematical problem at all: 'I did not verify my idea; I did not have the time since I was on the omnibus; I just resumed conversation, but I had an immediate feeling of absolute certainty'.

Discussing the apparently unconscious events which drive intuition, Poincaré concludes: 'You will be impressed by the appearance of sudden flashes of inspiration; the role of this intense unconscious activity in the mathematical inventions is definitely indisputable'.

close to it. When doing so one actually sees the conjecture in action. The immediate consequence is that we can confidently assume that there is an infinite number of periodic orbits as close as we like to any 'real' orbit, whether belonging to a celestial body or to a spacecraft. In the following chapters the advantages deriving from this approach will be discussed in detail. The aim is to share the feeling that periodic orbits are in fact a powerful means to open a breach in Poincarè's impregnable fortress.

3

Celestial waltz

A music that depicts nothing is a noise.
Jean D'Alembert

Playing the keys of a piano at random results in discordant sounds, but once the same notes are rearranged following harmonic laws, a melody is born. The ancient dream of celestial mechanics was to discover the 'music of the spheres' – the harmony hidden behind the motion of the planets as a sign of God's creation. Indeed, the development of astronomy showed that the Solar System is much more complex than previously thought. Besides the planets there are satellites, comets, asteroids, rings, and, as has been discovered in recent times, a whole new population of icy bodies orbiting beyond Neptune. Yet the motion of all these different celestial bodies is far from a random wandering in space, since in many cases they are 'tuned' in a sequence of harmonic chords. According to modern celestial mechanics one can say that the music of the heavens is a concert in which peculiar configurations – orbital resonances – play the main theme.

HEAVENLY RESONANCES

A sound is a vibration of a given frequency which propagates through the air. If it encounters an object on its way the vibration can be transferred, and the object starts oscillating at the same frequency. This explains why windows tremble to the point of cracking after a loud bang is heard. Such phenomenon is called *resonance*, involving the transfer of energy from one body to another. Resonances are widely used in building musical instruments: for example, the shape of a guitar has the effect of making acoustic waves properly resound in order to amplify the notes generated by the vibrating strings. Tuning among different instruments is also a matter of resonances, either obtained by means of electronic devices or musically educated human ears. The quality of the sound also depends upon resonances. The difference between an electronically generated single note and the richer and warmer sound played by a violin is due to the harmonics – the resonant vibrations with frequencies that are multiples (twice, three times...) or submultiples (half, third, three quarters...) of the original frequency.

On a more general and scientific ground, the frequency of a vibration is directly related to the period of the corresponding oscillation: the higher the frequency, the less time is needed to complete a whole oscillation.

As has been seen in the previous chapters, most celestial bodies travel, as a first approximation, along closed paths following their own footsteps with regular frequency. In this respect an orbital motion can be considered, from a mathematical point of view, the exact equivalent of an oscillation whose period is the timespan needed to complete an orbit. Celestial bodies become attuned when a particular relationship – resonance – is satisfied by the orbital periods: for example, when one period is the double the other, or one third, three quarters, and so on. In this case, perturbations act in a peculiar way, since symmetries are likely to characterise the orbital evolution of the system.

In order to introduce an example of orbital resonance, let us consider a system composed of Saturn and two of its moons, Mimas and Tethys. If we neglect the mutual attraction between the satellites, we obtain the two two-body subsystems Saturn–Mimas and Saturn–Tethys. Their unperturbed elliptical orbits have periods of revolution, say T_M and T_T. We know their values: T_M = 0.94 days and T_T = 1.88 days. It is easy to verify that 0.94 x 2 = 1.88, and therefore T_T has twice the value of T_M. Writing

$$2T_M = T_T \text{ or } 2T_M - T_T = 0$$

Mean motion resonances

The case of Mimas and Tethys is one of the most common orbital resonances occurring in our Solar System. They are also called *mean motion* resonances because they involve the periods of revolution of the celestial bodies. The mean motion is a quantity obtained by dividing a full 360-degrees angle by the period of revolution. In this way an averaged value of the angular velocity (in degrees per day) of the celestial body is obtained. As an example: if the period of revolution is 720 days, then the corresponding mean motion is 360:720 = $0^\circ.5$ per day. The Earth has a period of revolution of 365 days, which means that the mean motion of our planet is very close to 1° per day; that is, it spans roughly 1° of arc of its orbit each day.

In general, let us consider a system in which P_1 and P_2 are two celestial bodies orbiting a common central mass, while T_1 and T_2 are their periods of revolution. If there exist two integer numbers q and p such that

$$q\,T_1 = pT_2 \text{ or } T_1{:}T_2 = p{:}q$$

then the two bodies are said to be in a $p{:}q$ mean motion resonance. As discussed for the Mimas–Tethys case (where $q = 2$ and $p = 1$), a $p{:}q$ resonance implies that after q revolutions of P_1 around the central body, P_2 completes p revolutions. From a mathematical point of view, a mean motion resonance represents a peculiar choice between the periods of revolution, such that their ratio is not an arbitrary but a *rational* number.

is the usual way for celestial mechanics to explain that during the time needed by Mimas to complete two full revolutions around Saturn, Tethys has made exactly one revolution. Mimas and Tethys are then said to satisfy an orbital resonance 1:2.

The immediate consequence of the existence of a mean motion resonance is that the relative geometries among the celestial bodies involved repeat at every turn of the cycle. After two revolutions of Mimas around Saturn the relative positions of the satellite with respect to Tethys will be exactly the same during the next two revolutions, and then again and again. If we now leave the double two-body Saturn–Mimas and Saturn–Tethys approximation and restore perturbations, the existence of a resonance means that a particular geometry, such as the conjunction between the two satellites, always happens at the same point in their orbits. This has non-trivial consequences on the evolution of the system, since a conjunction represents the minimum distance between the two bodies and therefore allows the maximum direct gravitational attraction achievable (Figure 3.1).

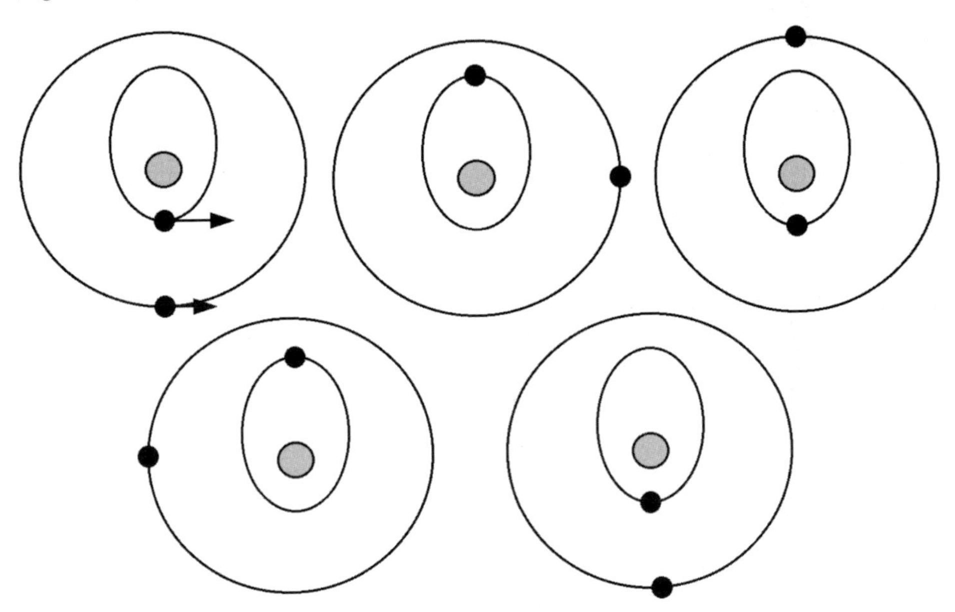

FIGURE 3.1. The orbital symmetries in the motion of two celestial bodies in the case of the 1:2 mean motion resonance. If one of the two orbits is eccentric and the two celestial bodies start at pericentric conjunction (top left), the resonance ensures that the two bodies will always revolve, keeping a safe distance between. The opposite effect happens for an initial apocentric conjunction, which implies that the minimum possible distance between the bodies has been reached and therefore will be reached again and again until direct perturbations cause the two bodies to exit from the resonance in a more or less dramatic way (a close encounter or a significant change in the orbital parameters).

Suppose that the inner orbit is not circular but has a significant eccentricity. If conjunctions occur when the inner body is at the apocentre, then the two bodies can approach dangerously close. On the other hand, if at conjunction the inner body is at pericentre then the two bodies are much farther apart and the existence of a resonance guarantees that this safe situation will not change in the future. In conclusion, resonances can act as an efficient protecting mechanism, and it is therefore not surprising that they are quite common in the Solar System.

COMMENSURABLE MOTIONS

The main belt is the region of space located between the orbits of Mars and Jupiter where the majority of the asteroid population resides. The first asteroid to be discovered, in 1801, was Ceres, and by now tens of thousands of asteroids have well-determined orbits. Because of the large variation in their periods of revolution – mostly from 3 to almost 8 years – the asteroid main belt is a perfect place for seeking mean motion resonances. One just needs to compute the ratio between the period of revolution of each asteroid (T_A) to that of Jupiter (T_J), and to then check the frequency of occurrence of resonances. Since the asteroids are inside the orbit of Jupiter, their period of revolution is always less than that of the planet, thus allowing only mean motion resonances characterised by submultiples of Jupiter's period, such as 1:2, 1:3, 1:4, 2:5, 2:3... and so on.

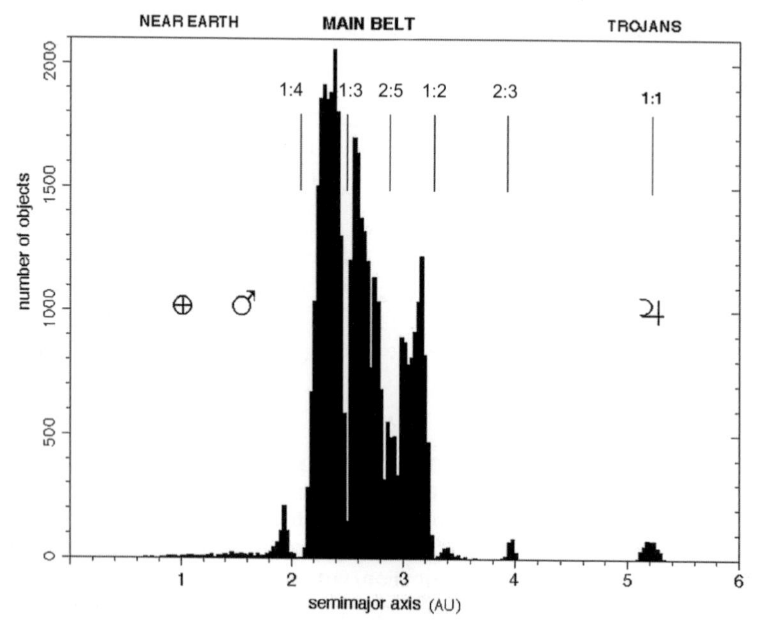

FIGURE 3.2. Distribution of the mean distances of the asteroids. The location of some mean motion resonances are marked, together with that of the planets, as denoted by their astronomical symbols.

An overall picture of the entire asteroid belt is obtained by using a *histogram*, where the number of asteroids having a period of revolution within a given range are plotted using columns with heights proportional to that number. The resulting diagram is shown in Figure 3.2, in which data have been ordered by the increasing mean distance of an asteroid orbit from the Sun (by Kepler's third law the mean distance is directly related to the period of revolution). This histogram represents the 'density' of asteroids as we recede from Mars and approach Jupiter.

It can be clearly seen that the asteroids are not uniformly spread within the main belt and that there are 'gaps' in the distribution – regions almost emptied of objects. More important is that these gaps occur for values of the orbital periods in mean motion resonance with Jupiter. In this particular case computations are easy, since the orbital period of Jupiter is very close to 12 years. The most prominent gaps are therefore found at values corresponding to $\frac{1}{4}\,T_J = 3$ years, $\frac{1}{3}\,T_J = 4$ years, $\frac{2}{5}\,T_J = 4.8$ years, and $\frac{1}{2}\,T_J = 6$ years. But it is not always so. Some asteroids show the opposite tendency to cluster at resonant values of the period of revolution, as in the case of $\frac{2}{3}T_J = 8$ years – a group known as Hilda's asteroids.

The opposite tendency to prefer certain resonances and avoid others is observed not only in the asteroid belt, but also in other densely populated systems such as planetary rings.

The Kirkwood gaps

Nowadays it is easy to access the Internet to obtain updated lists of known asteroids. Due to the progress in the sensitivity of telescopes and the development of automated search techniques, their number has been growing exponentially in the last decade, and now exceeds 100,000. This large number allows us to clearly see the 'holes' in the belt (Figure 3.2).

Life was not that easy when the gaps were first discovered by the American astronomer Daniel Kirkwood (1814–1895) in 1866. At that time there were only about 100 catalogued objects, and the convincing evidence came from the striking correspondence between the location of the shallow gaps and that of the mean motion resonances with Jupiter. These regions are now referred to in the scientific literature as the Kirkwood gaps – outstanding recognition for a man who began his career as a high-school teacher, quietly living in the State of Maryland. It was one of his students, with his intriguing questions on mathematics and astronomy, who pushed Kirkwood to resume university studies and to become a successful professional researcher.

Explaining the different behaviour of objects close to resonances is a longstanding problem for celestial mechanics. As far as the asteroid belt is concerned it took more than one century before a convincing dynamical mechanism was eventually proposed in the early 1980s by the American astronomer Jack Wisdom. Focusing on the 1:3 mean motion resonance – one of the deepest gaps in the belt – he discovered a subtle dynamical mechanism in which chaos plays a central role. Asteroids with an orbital period close to the 1:3 resonance, although apparently orbiting quite regularly, are subjected to sudden chaotic behaviour. The effect on the dynamics of these asteroids is the onset of large-amplitude variations in the orbital parameters, and in particular in the value of the eccentricity. Following these chaotic 'jumps' the orbit becomes so elongated that it crosses the orbit of Mars – a clear message for celestial mechanics. Sooner or later, on a relatively short astronomical timescale of a few tens of millions of years, the asteroid will have a close encounter with the Red Planet, resulting in a collision or in a dramatic change of its orbit. In either cases this leads to the removal of the asteroid from the main belt, its dynamics being no more controlled by Jupiter but by frequent encounters with the inner planets, including the Earth. This explains how the Kirkwood gaps formed and are continuously kept clear from any object falling into a dangerous resonance.

Mean motion resonances are easy to determine, since they involve the orbital paths of the celestial bodies. Yet they are not the only commensurabilities in the Solar System. With such commensurability we extend the concept of resonance to any relationship between the orbital parameters of two or more bodies that can be expressed by integer fractions. For example, 'secular resonances' are commensurabilities involving the motion of the line of the apsides (the line joining apocentre and pericentre) or of the nodes. In Chapter 1 it was shown that one of the effects of secular perturbations is a steady precession of the line of apsides. When the apsidal precessions of two or more planets are 'in tune', then a secular resonance occurs. In general they have a much longer period than mean motion resonances (hence the term 'secular', derived from the latin root 'century'), but their effect is equally important. As an example, a secular resonance with the apsidal precession of Jupiter runs through the whole asteroidal belt, shaping its bordering.

When dealing with commensurable motions the Poincaré maps (introduced in Chapter 2) are very useful in studying the complex relationship between stability and chaos, since they allow the possibility of finding 'islands' of stability amidst a sea of chaotic motions. Indeed, no general rule has yet been determined. Protecting celestial objects from dangerous dynamical configurations or throwing them into chaos depends upon the specific case taken into consideration. This is why mapping the 'geography of resonances' is one of the leading topics for modern celestial mechanics.

GREEKS AND TROJANS

Not all asteroids are strictly confined between the orbits of Mars and Jupiter. As shown in the histogram in Figure 3.2, some of them have the same mean distance and therefore the same period of revolution as Jupiter – a particular case known as a 1:1 resonance. If the orbits were circular and the resonance exactly satisfied, they should all be sharing the same orbit. This would be a dangerous configuration for a tiny asteroid affected by the gravitational pull of a massive

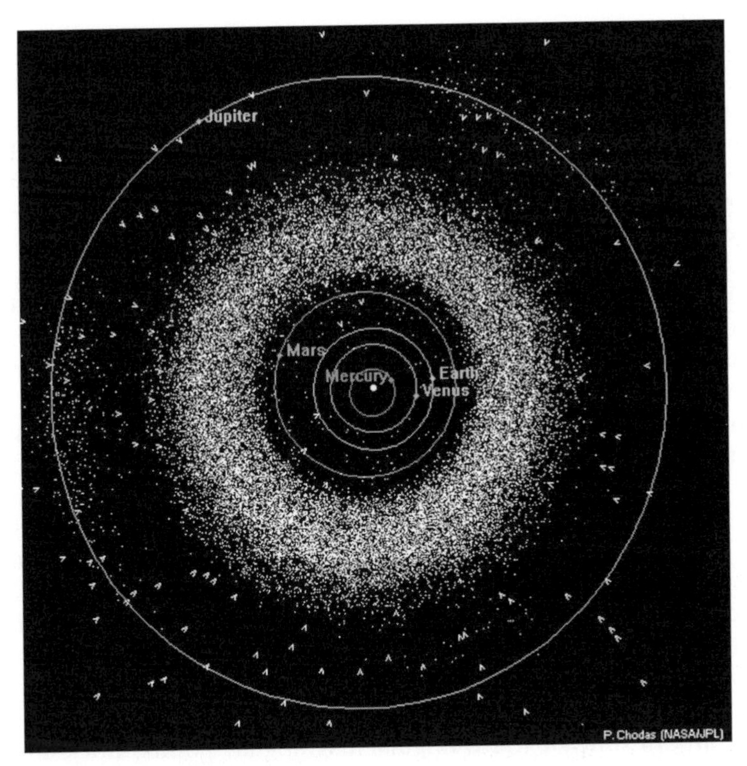

FIGURE 3.3. The doughnut-shaped main belt of asteroids occupies an extended region of space between the orbits of Mars and Jupiter. Outside it, the diffusion of objects toward the inner planetary region (the NEA population) and the clustering around the triangular Lagrangian points of Jupiter's orbit are clearly recognisable. Open triangles indicate the position of short-period comets. (Courtesy Minor Planet Center.)

planet. How could it survive over the age of the Solar System? The answer is obtained by looking at an 'aerial' view of our planetary system as in Figure 3.3, which shows the position of all known asteroids. Two groups of objects close to Jupiter's orbit can be recognised, one ahead and the other behind the planet. Although rather dispersed, the distributions are roughly centred around the triangular equilibrium Lagrangian points, L_4 and L_5, introduced in Chapter 2. Only the orbital patterns close to these special solutions of the three-body problem can guarantee the required long-term dynamical stability.

The two groups – known as the Greek camp and the Trojan camp – take their names from Homer's epic work the *Iliad*, and the outcome of the Trojan Wars is now depicted in the skies. The triumphant Greek army marches in front of the King of the Olympic Gods, while the Trojans sadly lag behind. Among the Greek asteroids, Achilles, Menelaus and Agamemnon can be found, while Aeneas, Priamus and Anchises are amidst the Trojans. Unfortunately, early hesitation in naming asteroids resulted in the placing of the Trojan hero Hector in the Greek camp and the Greek hero Patroclus in the midst of his Trojan enemies!

The 1:1 resonance has gained importance in celestial mechanics as a relatively stable region of space in which to search for small celestial bodies. Mars and Neptune are, to date, the only other planets with confirmed Trojans (it is now common practice to refer to celestial bodies located at one of the triangular Lagrangian points as 'Trojans'). In both cases two small bodies have been observed orbiting around L_4, but intensive sky surveys are underway and further discoveries are likely to occur. In this respect it is worthwhile noticing that it is apparent in Figure 3.3 that Jupiter's Trojans are not exactly located in L_4 and L_5, but are allowed to wander at much larger distances without escaping from the resonance. This explains why it is so difficult to find them associated with other planets, and extended observational surveys must be performed in order to cover a large portion of the sky.

Two of the largest satellites of Saturn have co-orbiting Trojans: Tethys (approximately 1,000 km in diameter) shares its orbit with the two 10-km-diameter moonlets Telesto and Calypso, located at the triangular equilibrium points of Tethys, 60 degrees ahead and behind Tethys, respectively. Similarly, Helene – a 32-km irregularly shaped body – orbits ahead of the satellite Dione, while Polydeuces – only 5 km wide – lags behind it.

The question has been posed whether the Earth itself has Trojans, small enough to have escaped direct observation. Extensive studies focused on the 1:1 commensurable motion with the Earth have shown that only three basic orbital patterns are possible: Trojan orbits are characterised by librations of various amplitude around one of the triangular Lagrangian points; horseshoe-shaped trajectories approach the planet on both sides, and the particle travels from one of the triangular Lagrangian point to the other; and chaotic regimes are easily recognisable because no regular pattern is followed. All these types of orbit can be connected by evolutionary paths, and horseshoe and Trojan orbits can be considered as transient stages of chaotic orbits. A small celestial body librating around a Lagrangian point can escape into a horseshoe configuration and *vice versa*. Chaotic orbits often lead to close encounters with the Earth, which may result either in trapping an object into Trojan or horseshoe orbits or in changing its semimajor axis to such an extent that a decoupling from the 1:1 resonance occurs. In conclusion it can be said that there is not sufficient dynamical ground for our planet to ensure rigorous stability around the triangular Lagrangian points, but that celestial bodies entering the 1:1 resonance with the Earth can be hosted for some time into relatively stable orbital patterns, as in the case of the small asteroid Cruithne discussed in the previous chapter.

This general picture has been substantiated with the advent of orbiting telescopes, which have detected a thin cloud of dust dispersed along the orbital path of our planet, temporarily trapped inside the 1:1 resonance. The light of the Sun is diffused by these particles, causing brighter regions to appear among the zodiacal constellations – a phenomenon known since ancient times as the *zodiacal light*, observable by the naked eye in particularly dark and clear skies.

FAMOUS AFFAIRS

Having at hand the tables reporting the orbital data of planets and satellites it is possible to search for mean motion resonances by simply dividing the periods of revolution of dynamically related celestial bodies and checking whether the result is close to an integer fraction (for example, ½ = 0.5, ⅓ = 0.333, ¼ = 0.25, ¾ = 0.75, and so on). As an example we take the value of the periods of revolution of Jupiter (T_J = 11.86 years) and Saturn (T_S = 29.46 years), and compute the ratio T_J/T_S = 0.403, which is remarkably close to the fraction $^2/_5$=0.4. The existence of a mean motion resonance between the two most massive planets of the Solar System is impressive, but it is not an isolated case and not the most accurate resonance! The 2:3 commensurability between Neptune and Pluto provides the basic dynamical mechanism for avoiding too close an approach between them. For the Saturn–Uranus (1:3) and Uranus–Neptune (1:2) pairs the corresponding resonant relations are satisfied with less accuracy.

The crowded satellite systems of the outer planets can be considered as scaled-down Solar Systems where resonances are also frequent. The four large satellites of Jupiter – Io, Europa, Ganymede and Callisto – named after some of the many lovers of Zeus, are dancing around Jupiter at the sound of a peculiar celestial tune: the pairs Io–Europa and Europa–Ganymede are both locked in a 1:2 mean motion resonance. As a consequence, the periods of revolution of Io and Ganymede are also commensurable, satisfying the ratio 1:4. This *multiple resonance* was first pointed out by the French astronomer Pierre Simon de Laplace, and again implies that certain geometries are either repeated or are carefully avoided. As an example, the Laplace resonance prevents the occurrence of triple conjunctions (the three satellites are aligned on the same side of the planet), which would give rise to a peak in their mutual perturbations (Figure 3.4). Similar commensurabilities also involve the uranian moons Miranda, Ariel, Umbriel and Titania.

The list of the Trojan and horseshoe configurations involving the moonlets and the satellites of Saturn into a 1:1 resonance is very long. Besides them, the pairs Mimas–Tethys and Enceladus–Dione are both in the 1:2 mean motion resonance; Titan and Hyperion satisfy a 3:4 commensurability; the ratio of the orbital periods of Titan and Iapetus is close to 1:5, and consequently Hyperion and Iapetus are also tied by a resonance (1:4). Again the stabilising effect of resonances can be inferred by studying in detail each specific case. An interesting example is that of Hyperion. This small irregularly shaped body is possibly the largest remnant of a catastrophic collision, as indicated by its chaotic rotation. Moreover, Hyperion has the 'bad luck' of orbiting just outside Titan (Figure 3.5), the planet-sized satellite (larger than Mercury) of Saturn – the only satellite of the Solar System surrounded by a dense atmosphere. Titan is closer to its home planet than Hyperion, and therefore moves faster, periodically overtaking Hyperion at conjunctions. When this happens the two bodies experience a minimum approach distance, while their mutual gravitational attraction reaches

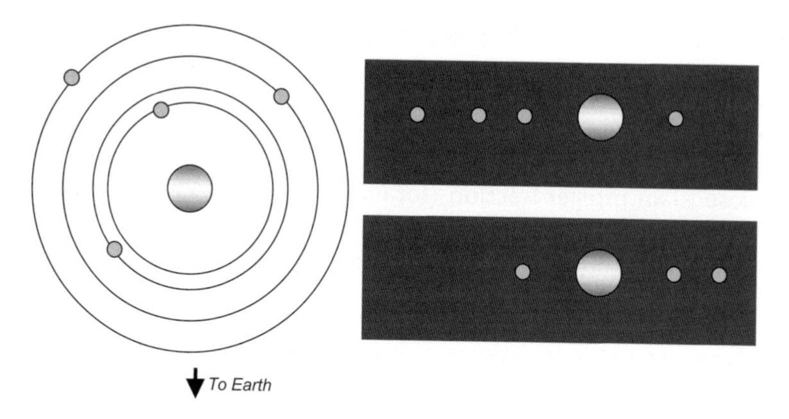

FIGURE 3.4. The Galileian satellites can be easily observed with small telescopes. The system (left) is almost coplanar to the ecliptic, and they therefore appear as bright starlets aligned on both sides of Jupiter (right). Their period of revolution is rather short, so that their relative position changes rapidly. Sometimes one or more satellites is hidden behind or passing in front of the planet, but among all possible combinations the Laplace resonance rules out the alignment of all of them on the same side of the planet. (Relative size is exaggerated in the diagrams)

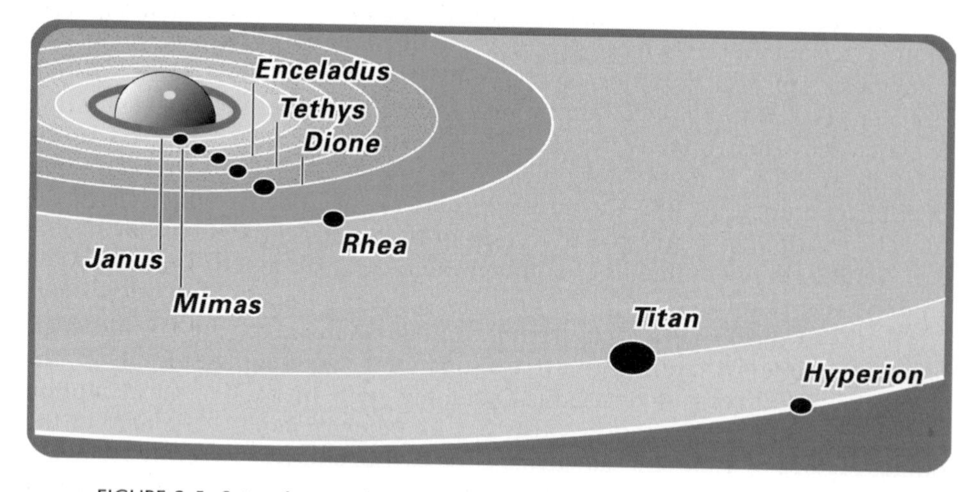

FIGURE 3.5. Saturn's crowded system of satellites is dominated by resonances.

a maximum. The sheltering action of the 3:4 mean motion resonance allows conjunctions to take place only when Hyperion is at the apocentre of its orbit, at the largest possible distance from Saturn and therefore also from Titan.

Exotic resonant mechanisms are also found in the Solar System. The existence of narrow arcs in Neptune's Adams ring has been explained as the observable effect of a high-order 42:43 mean motion resonance between the ring particles and the neighbouring satellite Galatea, which provides the necessary confinement of the particles in the ring (see Chapter 9). Even more intriguing are the commensurabilities involving the motion of the Moon, which leads to the Saros cycle (extensively discussed in Chapter 7).

During the last two decades, astronomical observations have shown that our planetary system is not unique. Indeed, many exoplanetary systems have been detected around other stars. Resonances are also often occurring among such exoplanets, even if at a first glance they appear of a completely different nature when compared to our Solar System. Apart from the dynamical interpretation in terms of long-term stability, resonant motion allows us to trust more confidently the complex and sometimes questionable procedures used to detect the presence of planets around a star (see Chapter 10). The very first reliable confirmation (after a couple of fake detections) of the existence of two planets orbiting around the pulsar PSR1257+12 was obtained in 1992, due to the existence of a 2:3 mean motion resonance which produced a characteristic signature of their orbital parameters.

THE CHANCE OF CHAOS

It has been repeatedly pointed out that resonances prevent the occurrence of dangerous configurations among the celestial bodies. A dynamical background to this generic statement is the ability to measure the stabilising effect of commensurable motions. Celestial mechanics is therefore used to investigate if and how resonances survive over astronomical timescales (billions of years), efficiently avoiding the onset of chaotic regimes.

In this respect an important result was obtained in 1954 by Archie E. Roy and Michael W. Ovenden, who estimated the probability that the frequency of the observed mean motion resonances could be due to pure chance. Their conclusion was that there are definitely too many commensurabilities in the Solar System to be treated as random events, thus calling for a dynamical explanation. Unfortunately this is not an easy task, since the way in which resonances act can be very different from one case to another, and the simultaneous occurrence of different types of resonance is often needed in order to provide stability.

Although general statements are quite risky, nevertheless the overabundance of resonant configurations can be interpreted as the result of an evolutionary process – a sort of 'natural selection' driven by celestial mechanics. Indeed, amongst the multitude of objects once crowding the Solar System, those which could rely on the strong sheltering action of one or more resonances had more chances to dynamically survive.

An outstanding example of this type involves Neptune and Pluto. The unusually large eccentricity of Pluto's orbit, if compared to the almost circular orbits of the other planets, implies a perihelion smaller than Neptune's distance from the Sun. Although at present the orbits do not cross due to their mutual inclination and angular parameters, they might in principle bring the two bodies dangerously close to each other. Such unpleasant configuration is prevented by the combined action of two different commensurabilities: the Neptune–Pluto 2:3 mean motion resonance and the secular precession of Pluto's perihelion, which avoids the minimum achievable distance between the two bodies ever being reached.

In this respect an intriguing topic consists in providing a modern answer to an old question. Can our planetary system can be considered globally resonant? Do the celestial bodies keep playing a cosmic chord – the long sought 'music of the spheres' hidden in their motion? As we have seen, the outer planets are the only ones remarkably close to mean motion resonances, even if the corresponding commensurabilities are not satisfied with high accuracy. The problem thus translates into understanding how far a resonance extends its action, which in turn requires the discovery of the meaning of being 'close' to a resonance. An interesting debate on this matter took place in the early 1960s. The two opponents were, once again, 'chance' against 'chaos', but no convincing arguments in favour of a global resonance have been put forward until now. We can only say that planets and satellites often prefer dancing at a simple and fascinating rhythm. Five turns of Jupiter are accompanied by two of Saturn, four turns of Titan correspond to three by Hyperion... and so forth. A celestial waltz in the skies.

COLOMBO'S EGG

In Bruce Murray's book *Journey Into Space* (1989), which describes the early pioneering years of the exploration of the Solar System, the following account appears in the section devoted to the Mariner–Venus–Mercury (MVM) mission:

> At about this time, MVM got another boost from the giant brain of Bepi Colombo. I barely knew this short, balding man, with one of the most engaging smiles in the world, when he showed up at an MVM science conference at Caltech in February 1970. Afterward he came up to speak to me.
>
> 'Dr. Murray, Dr. Murray', he said, 'before I return to Italy, there is something I must ask you. What should be the orbital period of the spacecraft about the Sun after the Mercury encounter? Can the spacecraft be made to come back?'
> 'Come back?'
> 'Yes, the spacecraft could return to Mercury'
> 'Are you sure?'
> 'Why don't you check?'
>
> He was right. After flying by Mercury, MVM would orbit the Sun with a period of revolution of 176 days, exactly twice that of Mercury's 88 days. With small manoeuvres the spacecraft could be made to return

FIGURE 3.6. Giuseppe 'Bepi' Colombo and the European Space Agency's BepiColombo twin spacecraft that will be sent for remote sensing the planet and probing of the magnetosphere.

> to Mercury's orbit every two mercurian years at the same time Mercury itself was there.
>
> Once again brain power had enhanced rocket power, as Bepi Colombo's fine mind grasped a profund reality that had escaped the rest of us.

This beautiful anecdote is a remarkable example of the benefits that can be achieved by the interdisciplinary nature of celestial mechanics. Giuseppe 'Bepi' Colombo (Bepi being the abbreviated form of Giuseppe) was an outstanding scientist who carried out seminal work on the rotation of Mercury and on commensurable motions in Saturn's system (Figure 3.6). His mind was used to dealing with resonances, and therefore he immediately recognised that the orbital period of the MVM spacecraft was very close to the 1:2 mean motion resonance with Mercury. Engineers thus determined that with minor fuel expenditures it was possible to perform repeated encounters with the planet, which eventually occurred on 29 March and 21 September 1974 and 16 March 1975 – once every two mercurian years.

Colombo's lesson has been thoroughly learned, and orbital resonances are now a powerful tool for space mission design. The tour of the Cassini mission to Saturn can be considered as a highly sophisticated exercise on commensurable motions (Figure 3.7). The many manoeuvres needed to continuously redirect the spacecraft in order to perform repeated close encounters with all the major satellites of the planet turned out to be highly demanding in terms of fuel consumption. They could not be achieved only by using the onboard propulsion system because of the huge propellant mass that needed to be carried from Earth. The solution was to exploit the large mass of Titan for gaining free gravity assists

Communication problems

The communities involved in studying perturbation theories, Solar System dynamics and spaceflight dynamics often suffer communication problems. But it is not easy to find a common ground for mathematicians, physicists and engineers working in different contexts and institutions such as universities, observatories, space agencies and the industry. Giuseppe Colombo (1920–1984) was one of the first to realise that modern celestial mechanics had grown larger than ever, spanning from the traditional speculative topics to the practical applications typical of the space age. More than that, he was convinced that spaceflight dynamics could bear original contributions to celestial mechanics, since man-made spacecraft can achieve dynamical configurations never observed among natural bodies. To him we owe the 'tethered satellite system' – an industrial application of a novel problem in celestial mechanics: studying the dynamics of a long tether orbiting around the Earth.

Nevertheless, Mercury was Colombo's favourite planet. Not only did he address a novel MVM mission design, but he also correctly explained the puzzling rotation of the planet. This is why one of the most challenging missions of the European Space Agency, which foresees two spacecraft orbiting the planet Mercury, has been named after him: BepiColombo.

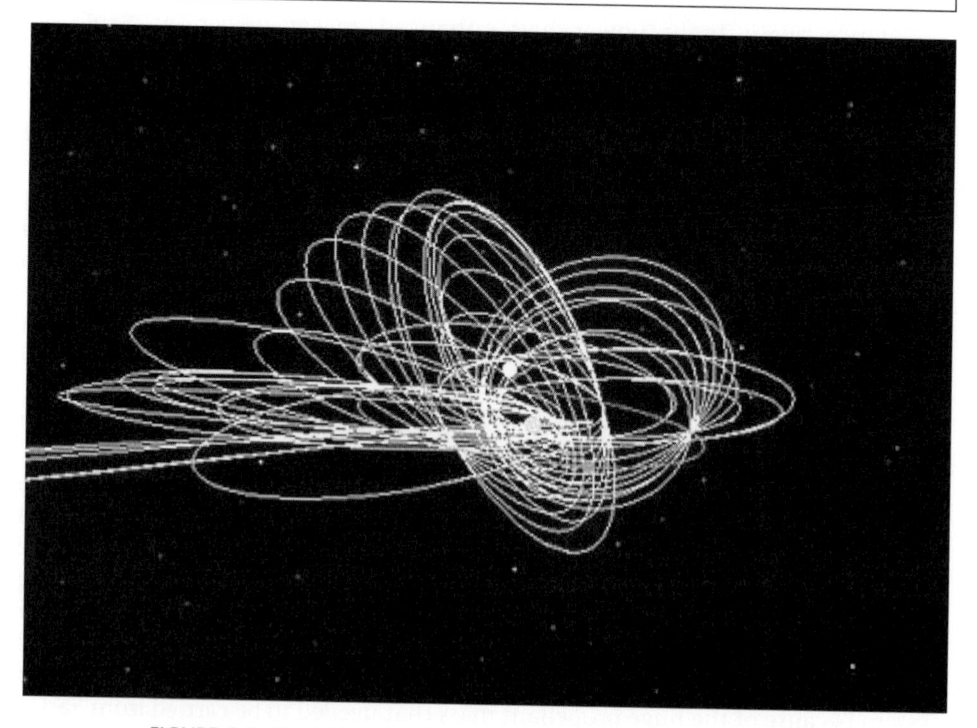

FIGURE 3.7. Cassini's tour of Saturn's system. (Courtesy NASA-JPL.)

and resonant orbits to 'sling' the satellite. In a 4-year nominal mission lifetime the Cassini spacecraft performs 44 fly-bys of Titan, jumping from one mean motion resonance to the other, chosen in advance for allowing the probe to visit the various target satellites. At the end of its high-tech 'waltz around the planet', the Cassini spacecraft will achieve a polar orbit allowing an unprecedented spectacular view from above Saturn's system.

4

Cosmic spinning tops

It is not because things are difficult that we do not dare;
It is because we do not dare that things are difficult.

Seneca the Elder

Poets, painters and musicians have often been inspired by the Moon, and their writings, paintings and songs dedicated to our satellite raise emotions and let the imagination run free. Yet every time we take a glimpse at the full Moon, it always shows us the same face, and this is even more evident when observing the lunar surface through a telescope. It is well known to astronomers that the Moon always shows the same hemisphere to the Earth and that this is due to the long-term gravitational interaction between the Moon and our planet. Tidal evolution is responsible for the peculiar relationship between the rotation of our satellite about its own axis and its orbital motion around the Earth, which causes the Moon never to turn its back on us. According to celestial mechanics, this is a classical example of a *spin–orbit resonance*.

SPIN AND ORBIT

The Moon is not the only known satellite showing the same hemisphere to its home planet, as most of the large satellites in the Solar System share the same attitude. This caught the attention of astronomers and defied celestial mechanics until a deeper understanding of the problem was provided by gravitation. Moreover, when interplanetary probes visiting the outer planets sent beautiful and detailed images of tens of other moons, it became clear that they were very different from each other: lunar-like surfaces battered by impact craters, highlands and mountains, but also great icy plains, grooved terrains, long cracks and deep valleys. Much to the surprise of planetologists, some of the new moons were 'alive': volcanoes erupting on the smooth surface of Io, a dense atmosphere surrounding Titan, geysers raising high on Triton, and possibly an ocean of liquid water hidden under the icy crust of Europa. Explaining the existence of these complex phenomena is not an easy task. Many branches of astronomy are involved, ranging from the origin and evolution of the Solar System to the internal structure of celestial bodies and planetary geology. Yet the coupling between spin and orbit is at the origin of them all.

The dynamics of an object of finite size is characterised by the rotation about its spin axis and by the orbital motion. The behaviour of a celestial body – a natural satellite revolving about its home planet or a planet about the Sun – can therefore be compared to that of a fast rotating spinning top describing an elliptical trajectory. But cosmic spinning tops also exhibit a close relationship between spin and orbit. A basic description of this behaviour involves only two quantities: the revolution period T_{rev} (the time needed to complete an orbit around the primary) and the rotation period T_{rot} (the time needed to complete a rotation about its own axis). The former is a more general definition of the 'year', while the latter applies to the duration of a 'day'. For the sake of simplicity in what follows it is assumed that the spin axis is perpendicular to the plane of the orbit (Figure 4.1). By using a pair of integer numbers p and q it is now possible to give the formal definition of a spin–orbit resonance of order p:q, occurring whenever the ratio of the periods satisfies

$$\frac{T_{rev}}{T_{rot}} = \frac{p}{q}$$

The simplest case is obtained when $p = q = 1$, which means that while completing one orbital revolution a celestial body also completes a whole rotation around its axis. When this happens the celestial body is said to be in a 1:1 or synchronous resonance (Figure 4.2). This is what characterises the motion of the Moon around the Earth, since astronomical measurements show that

$$T_{rev} = T_{rot} = 27.32 \text{ days}$$

Indeed, one lunar 'year' (a sidereal month) has the same length as a lunar 'day', and the Moon therefore turns around its axis at the same pace as it turns around our planet, thus always showing the same face.

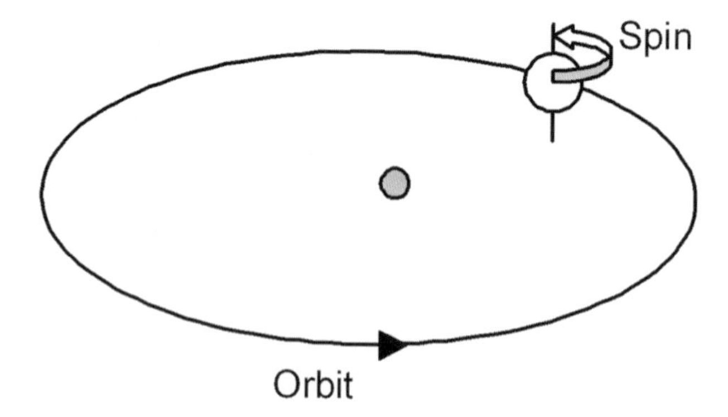

FIGURE 4.1. The coupling between the orbital motion of a celestial body and the rotation around its own axis has relevant consequences for celestial mechanics.

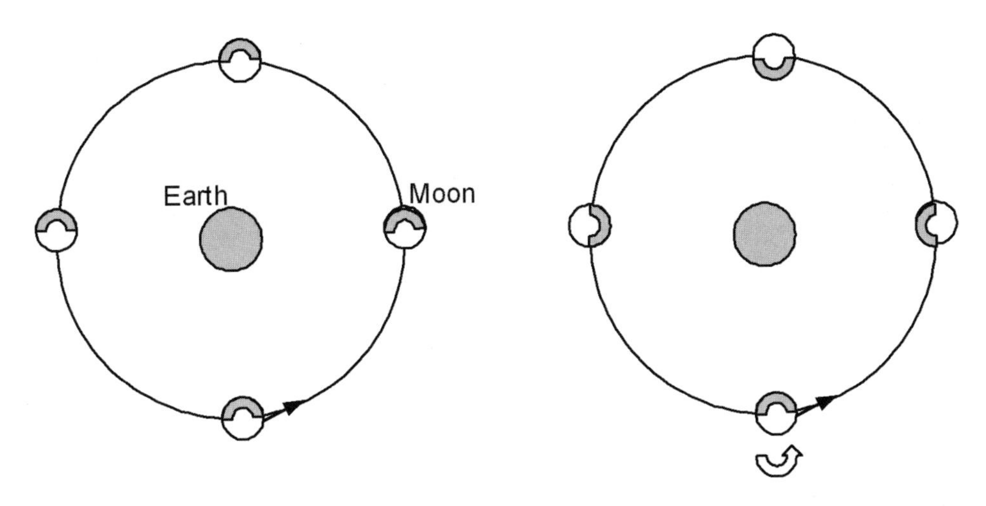

FIGURE 4.2. If the Moon were not rotating about its own axis (left) it would show all of itself to the Earth during a full revolution. Due to the synchronous rotation of the Moon, it always shows the same face as seen from Earth.

Another peculiar spin–orbit resonance occurs when $p = 3$ and $q = 2$, so that

$$\frac{T_{rev}}{T_{rot}} = \frac{3}{2}$$

which can be rewritten as

$$2T_{rev} = 3T_{rot}$$

This relationship better illustrates the resonance in action. During the time required for a celestial body to go through two revolutions, it also completes three rotations (Figure 4.3).

Following the same reasoning used in discussing mean motion resonances (see Chapter 3), the occurrence of spin–orbit resonances implies that certain geometrical configurations are repeated. Let us consider the case of a satellite with a 6-month rotation period and a revolution period of 9 months. If, at a given time, both the position of the satellite along the orbit and its orientation in space are recorded, 18 months (two revolutions) later the very same dynamical configuration is restored. Not only is the satellite back in the same orbital position, but it also shows the same hemisphere that faced the planet at the start of the resonant cycle, since an integer number of rotations around its axis (18/6 = 3) has been completed. What happened in the meantime? After one orbital period the satellite is back to the same point of the orbit, but after completing only 9/6 = 1.5 rotations, thus facing the opposite hemisphere to the planet.

Similar considerations hold when other fractions characterise the spin–orbit resonance, thus leading to the conclusion that only a synchronous resonance guarantees that the satellite hides half of its surface from its home planet at any time.

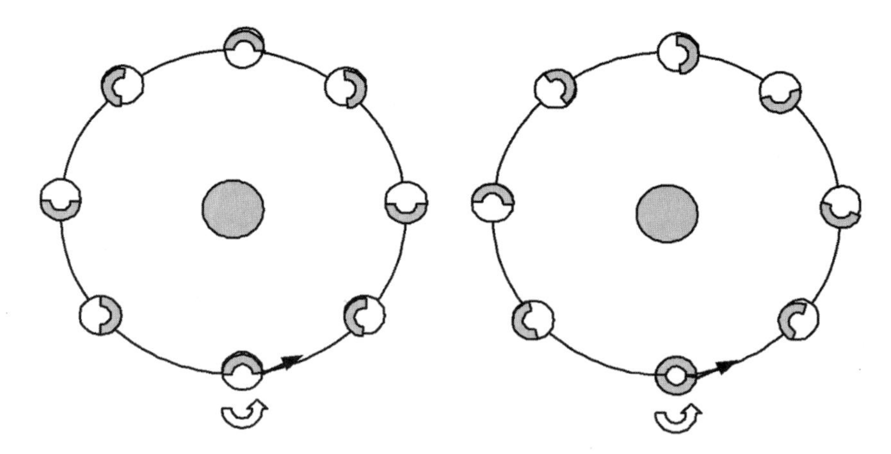

FIGURE 4.3. After a whole period of revolution a 2:1 spin–orbit resonance (left) brings back the celestial body to the same initial configuration, because it has completed exactly two rotations around its axis. Conversely, a 3:2 resonance implies that it has completed only 1.5 rotations over one revolution, thus requiring another orbit to be completed before the initial conditions are restored.

THE DARK SIDE OF THE MOON

A synchronous resonance allows us to see exactly only one hemisphere whenever the orbit is circular. A non-zero eccentricity does not change the orbital period, thus preserving the resonance, but it modifies the velocity along the orbit. This is the case for the Moon, which is subjected to consistent solar perturbations which are responsible for the small orbital eccentricity (e = 0.055). Applying Kepler's laws: at perigee the Moon travels slightly faster and therefore its orbital motion is not longer exactly synchronised with the rotation, but is a little ahead of it. At apogee the opposite situation occurs: the motion of the Moon is slower, thus losing angular speed with respect to the rotation. During one full revolution the two effects cancel out so that the resonance still holds, but these periodic velocity variations cause small amplitude oscillations of the Moon around the line joining its centre to the centre of the Earth. These are called *librational motions*. Something similar happens due to the non-zero inclination of the Moon and to the daily change of perspective for an Earth observer. Together these effects allow us to peer a little over the 'edge' of the lunar figure, so that overall we are able to see up to 60% of the Moon's surface.

Many legends and myths have evolved about what might be found on the 'dark side of the Moon' (which in reality never remains dark), including the possibility of exotic life forms. Lunik 3 – the first craft to successfully loop around the farside of the Moon – was launched by the Soviet Union in October 1959, and it sent back to the Earth images showing that the hidden face of our satellite is not much different from the one we know – a conclusion confirmed over the years by the many manned and unmanned missions that have reached our satellite.

Pulcinella on the Moon

John Herschel (1792–1871) was the son of William, the celebrated discoverer of Uranus. Unlike his father, who devoted his life to music and astronomy, he had many interests and made outstanding contributions in widely different fields of science, from mathematics to chemistry. He wrote a treatise on natural philosophy, and travelled to South Africa to survey the southern skies and to perform refined observations of Halley's comet on its 1835 return. His pioneering experiments on the early development of photography led him to become widely acknowledged all over the world. We owe to him the introduction of the word 'negative' for indicating the black-and-white reverse image appearing on photographic films and plates. Although John Herschel never held a professional position as an astronomer, in 1816 he took over his father's work in regularly observing the sky and published many scientific articles on astronomy.

It is therefore not surprising that Herschel became involved in one of the most celebrated media hoaxes of all time. In August 1853 an article was published on the front page of *The New York Sun,* in which an astounding series of astronomical discoveries were attributed to him. Among them there was the direct telescopic observation of intelligent life on the Moon in the form of winged human-like beings named *Vespertilio-Homo* ('man-bat'). Herschel was of course not even suspecting that his name was being used for selling thousands of copies of the newspaper, which never admitted that it was a hoax. Yet the story spread all over the world, and many people, even at university level, believed it true.

When the story of Herschel's alleged findings reached Naples, a new chapter of Pulcinella's saga was immediately added (see Figure 4.4). In it, Pulcinella visits the Astronomical Observatory of Capodimonte and asks the Director to translate from English the original article of the *Sun.* Not believing his ears upon knowing of the inhabitants of the Moon, he decides to reach our satellite by his own means. Pulcinella then builds a spaceship equipped with sails, wheels and a balloon, and off he sails to the Moon. Pulcinella's findings can be found illustrated in a series of refined and exhilarating prints that are now collector's items.

FIGURE 4.4. Pulcinella lifts off for the Moon.

After the magnificent enterprise of the lunar race which first took humans to the Moon in July 1969, interest in manned lunar exploration came to an abrupt end. Browsing through the list of space missions aimed at the Moon, it can be seen that there is a a large gap between 1974 and 1990, during which our satellite was not visited at all – not even by the tiniest automated spacecraft. In the late 1990s, interest was renewed with high-technology, low-cost exploration missions, bringing many opportunities for interesting discoveries. Craters permanently shadowed close to the lunar poles are now supposed to hide large reservoirs of water ice. When confirmed, this could dramatically change the scenario of a stable human settlement on the Moon, providing *in situ* resources such as water and oxygen-based fuel production.

The far side of the Moon (Figure 4.5) attracts astronomers, as they cannot imagine a better site for building telescopes – far from the disturbing action of an atmosphere, sheltered from the electromagnetic noise generated by our planet, and without the complexity of managing a low Earth-orbiting space telescope such as the Hubble Space Telescope. Yet the astronomer's dream of a permanent night allowing non-stop observation of the Universe cannot be fulfilled even by an observatory located on the dark side of the Moon. The reason is that the hidden face of our satellite is not dark at all! It suffices to recall that one lunar month is not only the revolution period of the Moon, but also, because of the 1:1 resonance, the length of a lunar 'day', so that during a lunar revolution around the Earth the entire surface is exposed to sunlight, much in the same way as happens to the Earth in a 24-hour terrestrial day (see Figure 4.2). In particular, the whole 'dark side' is fully illuminated at new Moon, when our satellite passes

FIGURE 4.5. On the way to Jupiter, upon returning for a gravity assist with the Earth, the Galileo spacecraft imaged the far side of the Moon (right) (Courtesy NASA/JPL). On the image at right, the dark regions are the borders of the large Oceanum Procellarum, which are also visible at the top of the image at left. The large concentric basin in the middle of the right-hand image is the 600-mile-diameter Orientale Basin.

through solar conjunction. These days the astronomically correct term 'far side of the Moon' is used (although Pink Floyd – the rock band who issued a record-selling album entitled *The Dark Side of the Moon* – will probably not agree).

TIDAL FRICTION

Even an essential list of Solar System bodies caught in synchronous resonance is rather long: the Moon and the two satellites of Mars, Phobos and Deimos; the four large Galileian satellites of Jupiter (Io, Europa, Ganymede and Callisto) and the small moon Amalthea; most of the satellites of Saturn (Mimas, Enceladus, Tethys, Dione, Rhea, Titan, Iapetus, Janus and Epimetheus); Uranus's moons Miranda, Ariel, Umbriel, Titania and Oberon; and Triton, the puzzling retrograde satellite of Neptune.

In addition, the case of Pluto is particularily interesting. In 1978 James Christy and Robert Harrington discovered that this small and unusual distant planet has a satellite, Charon. It was then found that not only the periods of revolution and rotation of Charon are the same (a classical synchronous resonance), but that they are equal to Pluto's rotation period. The immediate consequence is that Pluto always shows the same face to Charon, and the pair is said to be locked in *complete synchronous resonance*. If the Earth–Moon system were in the same situation, only people living on one hemisphere of our planet would be able to see our satellite in the sky; and those living on the 'wrong side' of the planet would always have the darkest nights.

A physical explanation of the overabundance of synchronous resonances in the Solar System is that they are the direct result of the long-term tidal interaction between bodies of finite dimensions. As discussed in Chapter 1, this interaction leads to a slowing down of the spin rate until a 1:1 resonance is achieved, because the tidal bulge is permanently locked towards the attracting body. Tides are experienced by both bodies, although with different intensity and timescales depending upon their relative size, mass and internal structure. The end state of a tidally interacting system is achieved when the revolution and the rotation periods of both bodies are equal; that is, when they revolve with their tidal bulges facing each other. This is the case in the Pluto–Charon system, in which both bodies are almost comparable in size and are trapped into a fully evolved 1:1 spin–orbit resonance. The Earth–Moon system has only partially evolved, while the outer planets have a mass much too large with respect to that of their satellites to be significantly affected by their tides.

The coupling between tidal interaction and orbital resonances might provide the answer to one of the most intriguing questions of modern planetary science. Why are the moons so different from each other? In order to fully understand the reason why the close-range images sent by the spacecraft visiting the satellite systems of the outer planets were so puzzling, it is necessary to refer to current views on how celestial bodies are born. The Solar System formed by the gravitational collapse of an intergalactic nebula composed of dust and gas, aggregating until kilometre-sized bodies (the *planetesimals*) appear. The runaway growth of planetesimals into protoplanets is driven by gravitational accretion, and is accompanied by intense heating. As a consequence, every planet at the end of its formation stage has inherited residual heat which depends upon the size and mass eventually reached by the planet. Over the age of the Solar System – five billion years – relatively small bodies such as Mercury and the Moon have already dissipated their primordial heat into space. Large planets such as Jupiter and Saturn are still hot enough to power large-scale atmospheric circulation, and they emit strong infrared thermal radiation. Earth-sized celestial bodies exhibit solid crusts and melted interiors that allow them to be geologically alive; and on their surfaces, volcanism and crustal dynamics is likely to occur (Figure 4.6). Yet they are just over the threshold. The surface of Mars (with half the diameter of the Earth) exhibits signs of a past activity. Large shield volcanoes, such as the gigantic Mons Olympus, are clearly recognisable, as is the 4,000-km long canyon system called Valles Marineris – all evidence of past significant geological activity on Mars.

Satellites formed much in the same way that the planets formed, but they are much smaller – on average, about 1,000 km. A few of them are larger than the Moon or even Mercury, but none are larger than Mars. According to the formation scenario described above they must have lost their primordial heat very early in their evolution, and should now exhibit the densely cratered surfaces resulting from billions of years of meteoritic bombardment. Why do medium-size celestial bodies such as Ganymede (with a radius of 2,631 km) and Io (with a radius of 1,822 km), or small bodies such as Tethys (with radius of 530 km) and Enceladus (with a radius of 250 km), clearly show recent or continuing geological activity? And where do they obtain the energy to sustain this activity?

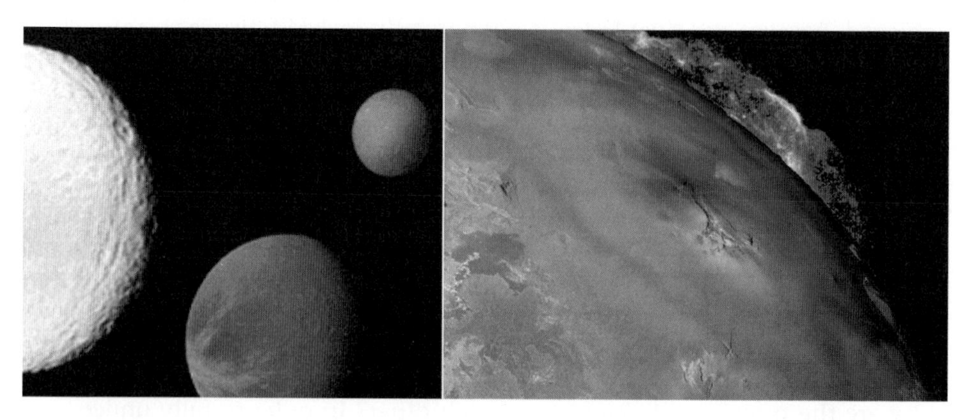

FIGURE 4.6. (Left) The surface of Saturn's satellites often exhibit signs of intense resurfacing events such as long cracks and bright plains. (Right) A high-resolution image of Pele – one of the most active volcanoes on Jupiter's satellite Io. (Courtesy NASA/JPL.)

Celestial mechanics provides an answer. When a satellite in synchronous spin–orbit resonance is also caught in a mean-motion resonance, the dynamical effect of the latter is to raise the orbital eccentricity. As a consequence the satellite undergoes periodic misalignments of its tidal bulge, similar to the librational motion of the Moon (described in the previous section). This generates internal friction because of the gravitational pull exerted on the bulge, tending to restore a synchronous rotation; and friction produces heat in quantities that depend upon the forces acting within the system. The Moon is far away from the Earth, and solid tides amount to just a few centimetres. But in the case of Io - which orbits around a planet 300 times more massive than the Earth and at a distance comparable to that of the Moon - the bulge rises up to 100 metres. It has therefore been computed that the perturbations on Io, due to the mean motion resonance and to the strong gravitational pull of Jupiter, produces an enormous amount of heat, able to melt most of the satellite mass. No wonder, then, that the images returned by the Voyager and Galileo missions show surfaces covered with many current volcanic eruptions - a phenomenon that has also intriguing geological implications because of the sulphur-rich composition of the lava flows and of the extreme conditions of the environment (low gravity and no atmosphere).

A striking prediction

In 1979 – only months before the first clear images of Jupiter's satellite system were returned by the Voyager spacecraft – a predictive article appeared in the international journal *Science*, in which the American researchers Stan Peale, Pat Cassen and Richard Reynolds evaluated the amount of tidal heat generated on Io as a result of the mean-motion resonance ruling the motion of the three innermost Galileian satellites. Eventually they came to the conclusion that Io could well possess an ocean of magma beneath a thin crust, thus producing widespread volcanism.

A few days after the first exciting encounter of the Voyager 1 spacecraft with Jupiter on March 1979, scientists were busily trying to discover exactly what was happening on the strange surface of Io, which appeared unlike any other body in the Solar System. The breakthrough came from a young member of the navigation team, Linda Morabito. Initially she had difficulty in matching the edge of Io's image with a circle (Figure 4.7); but upon enhancing the picture to investigate the problem she saw a 'bump' on Io's horizon showing the characteristic shape of a volcanic plume. It was one of the first spectacular confirmations of the predictions of celestial mechanics obtained during the space age.

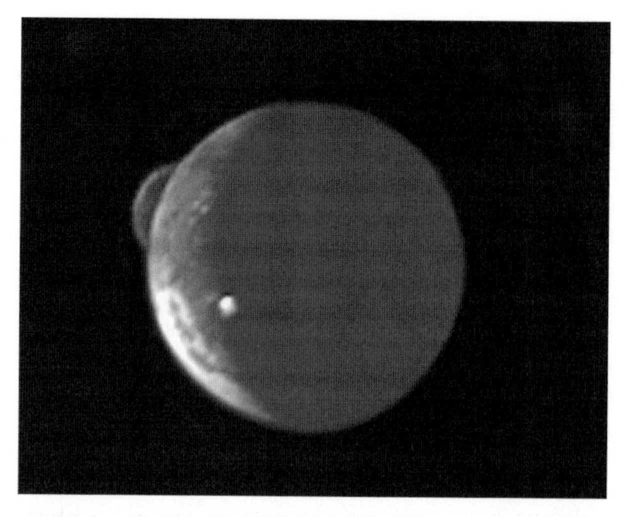

FIGURE 4.7. The historical image of the first alien volcanic eruption. (Courtesy NASA/JPL.)

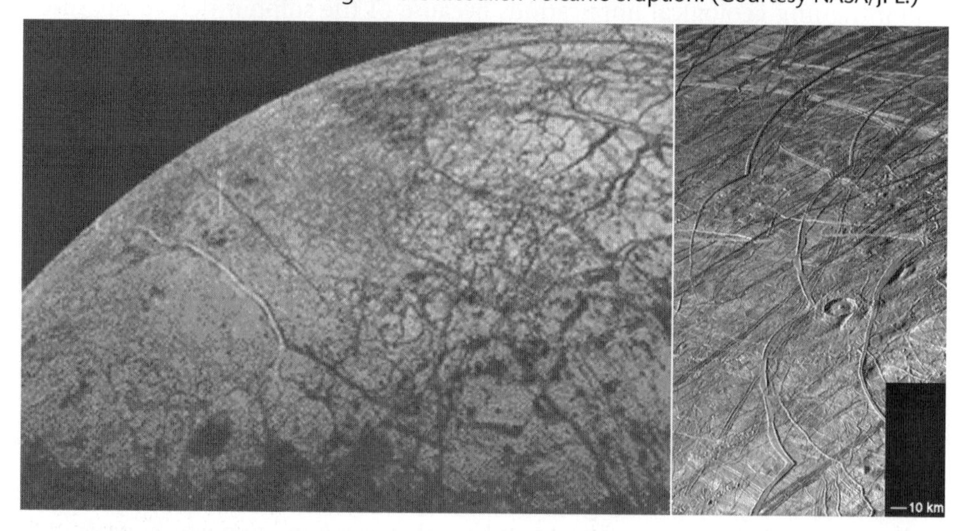

FIGURE 4.8. The cracked surface of Europa (left) and the cycloidal shaped fractures induced by tides (right). (Courtesy NASA/JPL.)

The effects of tidal friction on the other two Galileian satellites involved in the Laplace resonance have also been investigated. Another surprising result was found for Europa, which appears as a smooth icy world covered by a web of long streaks. Tidal heating of Europa is almost as efficient as on Io, leading to the conclusion that the satellite could maintain an ocean of liquid water beneath the frozen crust (Figure 4.8). The high-resolution images obtained by the Galileo spacecraft have provided partial confirmation by showing the break-up of large blocks of ice which appear to behave as huge ice rafts. But celestial mechanics has once again gone further. The American planetary scientist Richard Greenberg has (with his collaborators) pointed out that cycloidal cracks (a chain of arcs) observed on the surface of Europa can be attributed to the daily stress induced on the surface by tidal forces. There could therefore be regions which are periodically bathed in water flowing from beneath the surface and through the cracks – a scenario similar to the regular raising and lowering of the oceanic tides on our planet. A major consequence of this scenario is that it is well-known that shallow waters are potential niches for life, and organisms could have somehow evolved and adapted to the peculiar environment of Europa. The implications of these findings in the quest for life in the Solar System have pushed the major space agencies to initiate studies for a Europa orbiter – a spacecraft gravitationally bound to the satellite for continuous high-resolution monitoring, and possibly equipped with a surface element for performing *in situ* chemical and biological analyses.

After successful applications to Io and Europa, tidal friction has become a 'hot topic' when studying the geophysical evolution of synchronous satellites. The lightly grooved terrain on Ganymede and the bright icy surfaces of many moons around Saturn, Uranus and Neptune clearly show that impact craters have been covered by fresh flows of material coming from the inside. Now that the required energy source has been identified by celestial mechanics, planetary geology becomes responsible for a detailed description of cryovolcanism – a novel scientific term referring to low-temperature volcanic activity induced by tides.

GEOSTATIONARY SATELLITES

Many asteroids have satellites, the first of which was detected orbiting (45) Eugenia, while (246) Ida and its tiny moon Dactyl were imaged at high resolution

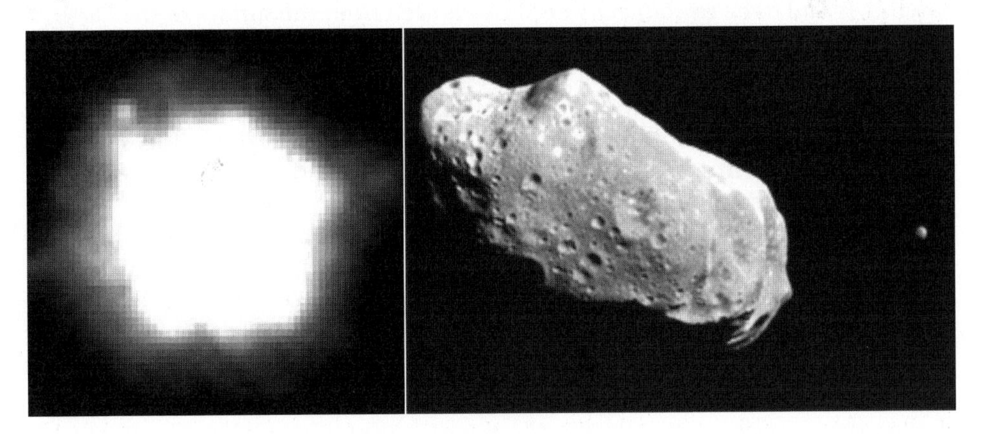

FIGURE 4.9. A satellite orbiting an asteroid was first observed from the ground in 1998 (left). The Galileo spacecraft discovered the tiny satellite Dactyl during its fly-by of asteroid Ida in 1993 (right). (Courtesy NASA/JPL.)

by the Galileo spacecraft in 1993, during a close encounter on the way to Jupiter (Figure 4.9). Binary systems are also becoming increasingly common among the transneptunian object population, and many others will probably be found in the near future. Yet at present the largest population of objects in complete synchronous spin–orbit resonance is not natural, but man-made: the geostationary satellites.

In order to send and to receive information from an artificial satellite (telephone, TV channels, telemetry, scientific data, and so on) a 'ground station' – such a large parabolic antenna or a domestic television set – must be kept in view of the satellite for as long as possible. The duration of the radio link depends upon the coupling between the orbital motion of the satellite and that of the Earth around its axis. An artificial satellite in a low Earth orbit (LEO, at 200–500 km altitude) has a short orbital period, typically of the order of 1.5 hours, thus moving faster than Earth's axial rotation. Henceforth the satellite's apparent motion in the sky is opposite to that of the Sun, and it rises in the west and sets on the eastern horizon. Visibility from the ground is reduced to roughly ten minutes. On dark clear nights and under proper illumination conditions, an artificial satellite can be recognised, even with the naked eye, as it appears as a tiny starlet moving fast across the sky.

Following Kepler's laws, the higher the altitude the longer the revolution period. Periods of visibility grow accordingly because of both the slower orbital speed and the longer orbital arc as seen from the surface of the Earth.

In 1945 the young astronomer Arthur C. Clarke – later to become a world-famous science fiction writer (the Stanley Kubrik film *2001: A Space Odyssey* is based on his book) studied the advantages of orbiting the Earth in a 1:1 spin–orbit resonance. In order to be *geosynchronous* an artificial satellite simply needs to have a 24-hour orbital period: the length of the day. If, in addition, the orbit is also circular and with no inclination with respect to the equator, then the satellite is said to be *geostationary* because, as seen from the surface of the Earth, it appears still in the sky. Yet the synchronous resonance between the Earth and the satellite is not complete unless the spacecraft is also spinning with the same 24-hour period. This condition is achieved (and maintained) by means of the onboard attitude control systems, and is essential for keeping the satellite's onboard instruments (cameras, parabolic antennae, and so on) always pointed toward the desired region on the surface of the Earth.

The first successful geostationary satellite, with an altitude of about 36,000 km, was deployed more than 30 years ago. Since then, hundreds of satellites, mostly used for telecommunication purposes, have crowded that region of space sharing almost identical orbits. As seen from outside, our planet would now appear surrounded by a *geostationary ring* of satellites positioned at different longitudes in order to cover the geographical areas below.

Space elevators

The idea of connecting the geostationary ring with the ground by means of a long pillar, thus exploiting the synchronous resonance for mechanically lifting satellites into space, is not new. In the early 1960s the Russian engineer Yuri Artsutanov wrote a visionary article entitled 'How to go in space by train'. Some years later, in 1975, the 'space elevator' (Figure 4.10) was 'rediscovered' by the American scientist Jerome Pearson. (Due to the political situation, Artustanov's paper did not have much diffusion in the Western world). The proposal was also visionary: 'The orbital tower: a launch system which exploits the rotation of the Earth'.

There is nothing wrong with this project. On the contrary, the possibility of going into space by climbing up a 'space ladder' is attractive, because it avoids managing the complex and dangerous rocket propulsion by substituting it with ordinary electrical engines. However, there is a technological problem concerning why space elevators cannot actually be built. There is no material that is at the same time sufficiently light and robust for realising a 36,000-km-long pillar. But in spite of these difficulties, space elevators are being studied because of their possible applications for exploration missions in gravity fields less intense than that of the Earth, such as the Moon and Mars.

FIGURE 4.10. A lunar space elevator reaching the L_1 Lagrangian point of the Earth–Moon system. (Courtesy Moon Base Conference.)

A PORTRAIT OF MERCURY

For a long time the planet Mercury – the planet closest to the Sun – has been thought to be in a 1:1 spin–orbit resonance. It was a reasonable guess in the light of tidal evolution: the enormous solar tides raised on its surface could spin it down to synchronous rotation in a rather short time. Showing always the same face to the Sun has extreme consequences: a never-ending burning day on one hemisphere and the coldest eternal night on the other. If this were true, half of Mercury's surface would have an extremely high temperature, while the other would be freezing. There was general agreement in the astronomical community until a few decades ago, when an excess of heat emission from the dark side of Mercury was detected by radio observations. It was puzzling evidence, but the belief that the planet was trapped in a synchronous resonance was so strong that many scientists tried to dismiss the observations by assuming the existence of a thin atmosphere responsible for heat diffusion all around the planet. More detailed measurements from the Earth and the direct exploration of the planet by the Mariner–Venus–Mercury (MVM) mission in 1974–75 solved the riddle, showing that the spin–orbit coupling of Mercury follows a 3:2 spin–orbit resonance.

From the presently known values of the mercurian periods of revolution (T_{rev} = 87.97 days) and of rotation (T_{rot} = 58.65 days), a simple computation yields T_{rev}/T_{rot} = 1.4999, whereas $\frac{3}{2}$ = 1.5 and the resonance is satisfied with a high degree of accuracy. Every two orbits around the Sun (87.97 × 2 = 175.94 days), Mercury completes three rotations around its axis (58.65 × 3 = 175.95 days), so that during one full resonance cycle the whole surface of Mercury is exposed to the Sun, thus explaining the observed excess heat emanating from the dark side of the planet.

Even before the experimental evidence appeared, such behaviour was first proposed in 1966 by Giuseppe Colombo, who showed that in the case of Mercury the 3:2 spin–orbit resonance is a resonance as stable as the synchronous rotation. Almost ten years later confirmation came with the MVM mission, which also profited by Colombo's familiarity with commensurable motions. As we have seen in Chapter 3, the orbital period of the MVM spacecraft was remarkably close to the 1:2 mean motion resonance with Mercury – which allowed the performance of three encounters with the planet at the price of one.

There is, however, a drawback to this success story. Notwithstanding the increased number of encounters, MVM was able to image only about 50% of the surface of Mercury. Due to the coupling between the 1:2 mean motion and the 3:2 spin–orbit resonance, the planet was always encountered at the same position along its orbit and therefore in the same rotational state, with the same hemisphere facing the Sun. From the point of view of mission analysis, the two resonances were fighting one against the other. Resonances are particularly difficult – and they protected Mercury's privacy from the electronic eyes of the MVM probe.

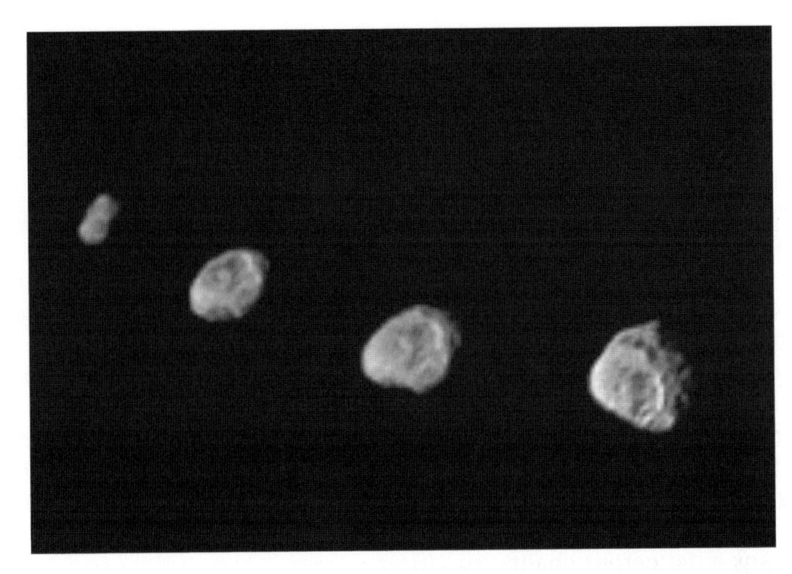

FIGURE 4.11. The peculiar shape of Hyperion as seen in four different images obtained by Voyager 1. (Courtesy NASA/JPL.)

TIDY CHAOS

Among the regular satellites of Saturn, Hyperion is characterised by a remarkable irregular shape (Figure 4.11). In 1980 and 1981 the Voyager spacecraft returned a spectacular sequence of images showing a 'hamburger-like' celestial body in a definitely chaotic state of rotation – a crazy spinning top in the sky. This peculiar motion can be understood by invoking several concurring factors: the non-spherical shape of Hyperion (its dimensions are approximately 164 × 130 × 107 km); its unusual origin (possibly a remnant of a catastrophic collision or a captured body); the significant orbital eccentricity; and the strong perturbations exerted by its neighbour, the large satellite Titan.

Synchronous resonance and chaos represent extreme cases of spin–orbit interaction. For a better description of the whole phenomenon we can exploit the analogy with the motion of a pendulum, because both – a commensurable motion and the dynamical behaviour of a pendulum – are described by similar equations. Let us take a 'perfect' pendulum: no dissipative forces such as air resistance, or friction at the anchorage point, are present. When the pendulum is at rest and no external forces are acting, it remains indefinitely in its equilibrium position. A small external force will cause oscillations to begin, with amplitude dependent upon the intensity of the initial perturbation. Small displacements from the equilibrium position are called *librations*. If the perturbing force is so strong that the pendulum starts to revolve around the anchorage point, the resulting motion is called *circulation*.

Returning to celestial mechanics, the equilibrium position of the pendulum corresponds to exact synchronous resonance, when a satellite always shows the same face to its host planet. A small displacement from the exact resonance triggers librational motions, like those mentioned in the case of the Moon. The satellite oscillates slightly back and forth with respect to the line joining it to the planet. If the amplitude of these librations exceeds a certain limiting value, the onset of a circulation regime brings the satellite to free rotation, and it shows the whole surface to the planet. Chaos is a mixture of all these orbital regimes. Outside the protecting action of a resonance, the motion of the pendulum becomes a complex and unpredictable sequence of motions such as circulations and librations of any amplitude.

The interplay of ordered and chaotic behaviours can be also described on more familiar ground. We all know that a safe see-saw ride requires moderate pushes while keeping the same pace as the oscillations of the see-saw. In doing so we ensure that we remain in the stable domains of resonances and librations. Pushing too strong would lead to frightening circulations, while decoupling the frequency of a push from that of the oscillations would result in loss of control of the see-saw: a dangerous chaotic situation.

The high number of occurrences of synchronous resonances indicates that celestial bodies appear to have learned quite well how to ride the cosmic see-saw of gravitation.

THE OBLIQUITY OF THE PLANETS

It has so far been assumed that the spin axis of a celestial body is perpendicular to the orbital plane. But reality can be very different, as shown in Table 4.1, where the actual inclination of the spin axes of the planets – the *obliquity* – is listed (see also Figure 4.12). The obliquity of the planets ranges from a few degrees, as in the case of Jupiter, to about 90° for Uranus, which appears to be tumbling along its orbit. Venus has a retrograde rotation – a unique case among the planets.

There are two major consequences of the non-zero obliquity of the planets. One is of course relevant for celestial mechanics, which has been asked to explain how it affects the dynamics of a rotating spinning top in the sky. The other consequence has profound implications on the physical evolution of a planet, as it affects the geometry with which the solar radiation reaches the different regions on its surface. The very same existence of the seasonal cycle on our planet and of the annual variation of the length of the daylight are striking evidence of both the non-zero obliquity of the Earth and of the impact on natural phenomena: from the development of ecosystems to human social behaviour.

Moreover, planetary spin axes are not fixed in space, but move slowly, following a conical path perpendicular to the orbital plane. This *precessional motion* can be visualised as being similar to the whirling oscillation of a spinning top at the end of the run. In the short term it modifies, to a minor extent (the

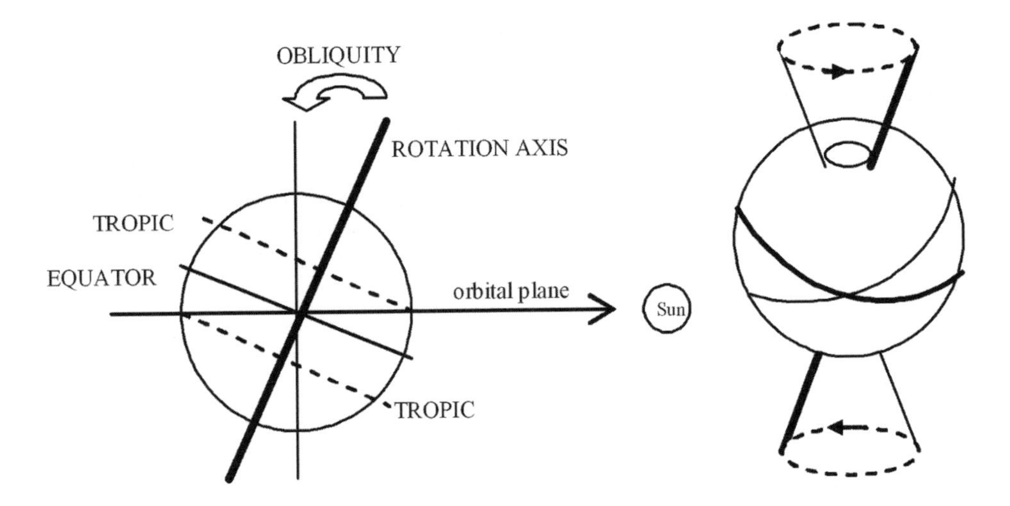

FIGURE 4.12. The non-zero inclination of a planet's rotation axis is called *obliquity* (left). The precessional motion of a planet is similar to the one typical of a spinning top (right).

TABLE 4.1. The obliquity of the planets.

Planet	Obliquity (°)
Mercury	47
Venus	179
Earth	23
Mars	23
Jupiter	3
Saturn	27
Uranus	98
Neptune	29

'wobbling' of the spin axis), the value of the obliquity, but it rather changes the orientation of the spin axis in space. If it were not so, life on Earth would be endangered, as a change of only 1° in the obliquity would have catastrophic effects on the climate on a global scale.

Planetary precession acts on astronomical timescales (the spin axis of the Earth takes about 26,000 years to complete a full cycle), and is determined by the coupling of the oblateness of the celestial body with the gravitational forces acting on it. Thus celestial mechanics is faced with a problem which deserves maximum attention. In the long term will chaos also appear in this type of motion? And was it already present at some time in the past?

In order to detect possible long-term variations of the obliquity of the planets it is necessary to follow their dynamical evolution over timescales of at least several tens of millions of years. The consequent growth of computational errors when integrating the equations of motion for long timespans, especially when dealing with chaotic dynamics, has for a long time presented difficulties when approaching this problem. The recent development of novel mathematical techniques such as *frequency analysis* has allowed the French astronomer Jacques Laskar to perform extensive and reliable numerical experiments demonstrating that the rotation of the outer planets is stable (although the anomalous obliquity of Uranus still awaits explanation), while the results obtained for the inner planets are surprising.

The inclination of the spin axis of Mercury has presumably wandered chaotically in the past until it was caught in 3:2 resonance. Venus has a similar chaotic history, which has led to the complete inversion of its rotation axis, as is observed today – an explanation that rules out previous hypotheses involving cosmic catastrophes which, in passing, had to be of incredible proportions to be able to turn an entire planet upside down!

This is good news for our planet: the obliquity of the Earth is definitely stable, and a chaotic regime is possible only if the obliquity exceeds 60° – a figure far enough from the present value (23°) and not likely to be achieved. The presence of the Moon – an unusually large satellite compared to the mass of its home planet – is largely responsible for this situation, since its gravitational pull has a long-term stabilising effect. This provides additional scientific ground concerning the fundamental contribution of the Moon to the development of life on Earth. Our fortunate position is even more evident when looking at the Earth's brother planet. Mars has an obliquity remarkably close to that of Earth; but this similarity seems to be accidental, as according to Laskar's computations it is deeply inside a chaotic region. The climatic variations triggered by an ever-changing obliquity must have been dramatic for Mars, particularly in shaping the rocky desert covering the whole martian globe that we see today. It must not have been always so. Thanks to the many space missions orbiting the planet the areography of Mars is well known, and the past presence of water is now considered as certain. This evidence appears in the form of typical geological features such as dry river-like channels, smooth terrain at the bottom of impact craters indicating the presence of ancient lakes, and lowlands extending over most of the northern hemisphere of the planet – possibly the sea floor of a great ocean reaching up to the pole.

RAIDERS OF THE LOST EQUINOXES

It is everyday experience that the length of daylight is not constant during the year. It reaches a maximum at the beginning of summer and a minimum when winter arrives, while day and night are equally shared at the beginning of autumn and spring. These events are called, respectively, *solstices* and *equinoxes*, and they mark the cycles of the seasons. This is not mere coincidence. On a global scale the climate is determined by two basic factors: the inclination with which the solar

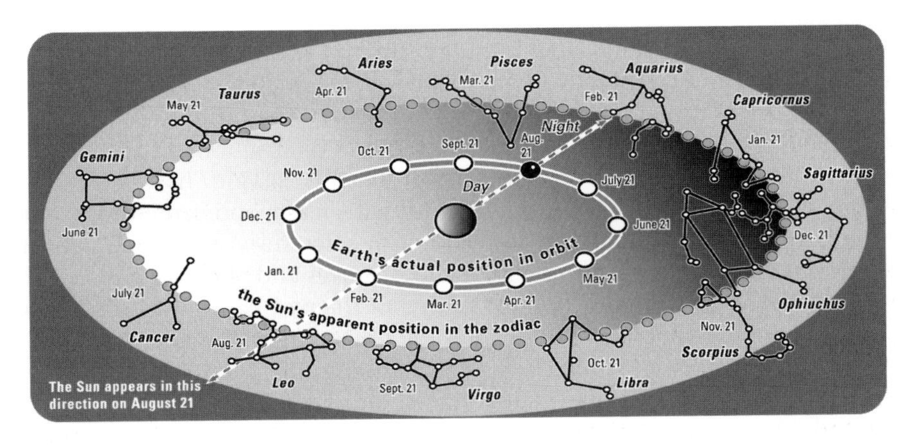

FIGURE 4.13. The relationship between the zodiacal constellations and the orbital motion of the Earth.

radiation meets the ground (at 90° the incident light has less atmosphere to travel through before reaching the surface, which in turn results in less absorption), and the duration of the exposure to the warmth of the Sun. Both effects are more sensitive to the non-zero obliquity of the Earth rather than to the slightly varying distance of our planet from the Sun because of the non-zero orbital eccentricity.

While travelling along its annual orbital path our inclined planet offers changing perspectives to the incoming solar rays, which can be assumed parallel to the ecliptic. At the summer solstice the north pole and the northern circumpolar regions are exposed to continuous daylight (Figure 4.13), and toward the winter solstice the days become shorter and the midday Sun is lower on the horizon. At the same time, in the southern hemisphere the situation is reversed and summer arrives.

The existence of equinoxes and solstices can be considered as one of the first astronomical observations made by mankind – a knowledge that dates back to ancient civilisations.

In the second century BC the Greek astronomer Hipparchus (*c.*185–125 BC) first noticed that the orientation of the celestial sphere (the positions of constellations in the night sky at a given time) was slowly changing over the years. This phenomenon – the *precession of the equinoxes* – is the observable effect of the precessional motion of the terrestrial axis mentioned in the previous section. The steady motion of the Earth's axis on the celestial sphere introduces a growing time displacement of the seasonal change, thus decoupling heavens and Earth. If we had a time machine we would see that in 13,000 years from now (half of the precessional cycle) the constellations visible today during summer nights will appear in the winter sky. The same applies when travelling backwards in time: 13,000 years ago our winter constellations appeared during the summer.

This is why the precession of the equinoxes is very helpful for archaeologists when dating ancient sites, as it can be considered a celestial clock. On its quadrant, instead of hours and minutes, time is measured in centuries and millennia. In particular there is an observable event which allows us to trace back in time the orientation of the spin axis of our planet: the position of the Sun on the celestial sphere (the zodiacal constellation that it occupies) at the time of the spring equinox (Figure 4.13). The arrival of this season has been always welcomed by religious or pagan populations with feasts and ceremonies, and it is therefore not difficult to find astronomical references in ancient writings, inscriptions and paintings. Moreover, temples, graves and religious buildings can be considered as 'astronomical observatories' of the past, since they often report in their construction plans the geometries observed in the sky. Well-known examples are the ancient stone circle of Stonehenge, the pyramids of Gizeh, and the highly sophisticated meridians traced on the floors of churches and cathedrals. Yet some of those geometries are not valid today: Earth and sky might no longer be aligned, because of the long-term drift of the constellations. Travelling backward in time until the disagreement disappears is the domain of archeoastronomy, which exploits celestial mechanics to place, in the proper time frame, the archaeological and historical findings. Precession of the equinoxes also causes the pole star to change over time (Figure 4.14).

IS THE LAND OF ATLANTIS REALLY LOST?

The precession of the equinoxes is so slow and its observable effect so small in the course of a lifetime that it might be wondered how Hipparchus managed to discover it. A fascinating hypothesis was first proposed by the Irish historian J.V. Luce, relying on the researches carried out by the Greek archaeologist Spiridon Marinatos and later refined by others.

It began in ancient Greece in the fourth century BC, when the King of Macedonia asked his favourite poet, Aratus of Soli (315–240 BC), to translate into more friendly words the treatise on astronomy by Eudoxus of Cnido (409–356 BC). Eudoxus had written it upon returning from a long voyage to Egypt, where he had learned of an ancient civilisation which dominated the eastern Mediterranean and had advanced astronomical knowledge. The Egyptian astrologers told him that this great civilisation disappeared abruptly, from one day to another, and that all that remained was a globe made of stone which they received as a gift and was very useful for navigation. Eventually the Egyptians gave the globe to Eudoxus, who took it to Greece. He certainly used it when writing his book, because it had the constellations and some important astronomical references engraved on it, although he did not check whether it was sufficiently accurate.

We do not know if Aratus succeeded in making celestial mechanics more appealing to the King of Macedonia by rewriting Eudoxus's treaty; but what we *do* know is that the globe and Eudoxus's book were lost, and we have Aratus's poem, which survived. Upon reading it two major points can be highlighted. First, the constellations on the globe bore the names still used today; and

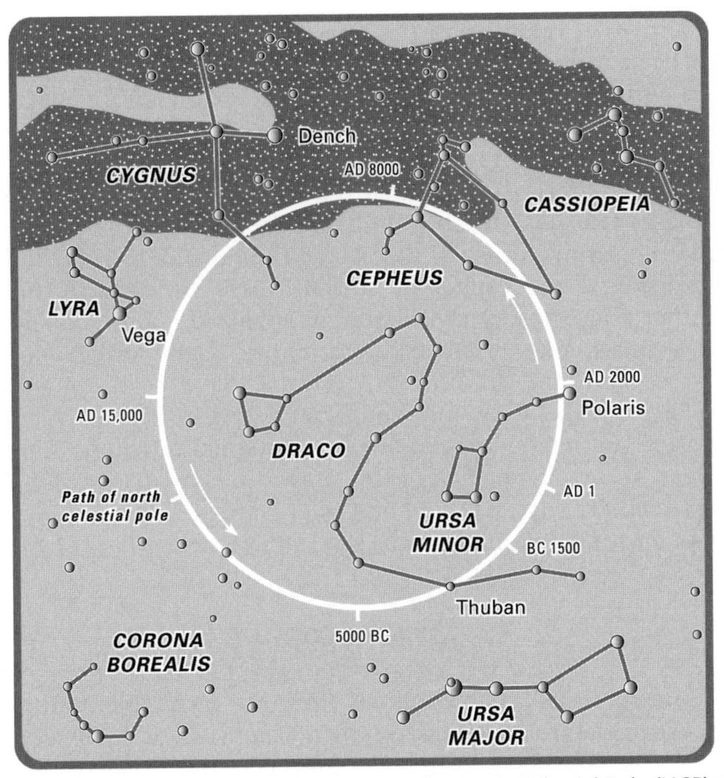

FIGURE 4.14. Precession of the equinoxes causes the North Celestial Pole (NCP) to move slowly in a circle around the North Ecliptic Pole, causing the pole star to change over time. In ancient Egyptian times, for example, the NCP was close to Thuban in Draco (the Dragon). Today, the star Polaris in Ursa Minor (the Little Bear) is the nearest naked eye star to the NCP. Around 1500 BC there was no bright pole star.

secondly, some very bright stars that Hipparchus could not see were engraved on the globe, while some others easily visible at the time of Hipparchus were not reported at all.

It was probably this last observation that put Hipparchus on the road to discovering the precession of the equinoxes. The globe effectively recorded the sky as it had appeared much earlier, thus representing precious data on the past dynamical history of our planet.

But how old is the globe? To answer this question we must investigate the origin of the constellations, because it was probably the ancient civilisation that produced the globe that first grouped the stars in the sky and named them. Archeoastronomers have used a neat trick to this end, by compiling a list of 'wrong' astronomical statements extracted from Aratus's poem and then using a planetarium to go back in time until the precession of the equinoxes makes them become true. When this is done, the constellation makers are found to have lived about 1800 BC, at a latitude of approximately 36°. This result is in remarkable agreement with what is known of the Minoan civilisation, which developed on the island of Crete at about that time and ruled the Mediterranean with a powerful fleet. This is also a reminder of a classical mythological story. Trapped inside a maze under the royal palace of Crete, there lived the bull-headed monster, the Minotaur; and from Crete, Icarus made the first human flight.

Besides these myths, the existence of such an evolved civilisation was demonstrated as late as at the beginning of the twentieth century, when the British archaeologist Arthur Evans discovered the ruins of the magnificent royal palace of Knossos - the capital of ancient Crete. Subsequent studies have confirmed that the Minoan civilisation disappeared quite suddenly - most probably as a consequence of the explosion of the neighbouring volcanic island of Thera (now Santorini) in 1450 BC. It must have been an unimaginable catastrophe: earthquakes, tsunamis, and ashes falling from the sky and covering the whole island of Crete to the point of making agriculture impossible for many years to come. The remembrance of this dramatic event has possibly survived. After being recorded by the Greek philosopher Plato (428–348 BC), it reached us as one of the most fascinating legends of all time: the lost land of Atlantis.

ASTRONOLOGY

The long journey which has brought us from the most recent studies on the obliquity of the planets to unveiling the origin of constellations has allowed a brief glimpse of celestial mechanics in an historical perspective. In this respect it can be said that the precession of the equinoxes marks the borderline separating modern astronomy from ancient astrology. Trying to understand why and when their paths started to diverge is an intriguing exercise of applied celestial mechanics.

Everyone knows the answer to the question: 'What is your zodiacal sign?' Indeed, one year is divided into twelve intervals of time named after the zodiacal constellations (Aries, Taurus, Gemini, and so on), and it is easy to check which sign fits a given birth date. This simple task has many astronomical implications. The relationship of the twelve time intervals and the zodiacal signs reflects a well-defined astronomical event: the passage of the Sun through the zodiacal constellations.

The stars are always present in the sky, even if the sunlight diffused by the atmosphere, painting the sky blue, does not allow us to distinguish their faint glittering during daytime. From space the Sun appears like a sharp disk on a black background dotted by stars, and from there it would be easy to observe that during one year our star follows a regular path in the sky. It is an apparent motion due to the change of perspective with respect to the celestial sphere as the Earth proceeds along its orbit (Figure 4.13). Astronomy and astrology agree on that.

Nevertheless, as we have seen, the precession of the equinoxes introduces a small drift: every year the turn of the seasons arrives earlier, astronomically speaking, by about 20 minutes. If not properly accounted for in the calendar, as

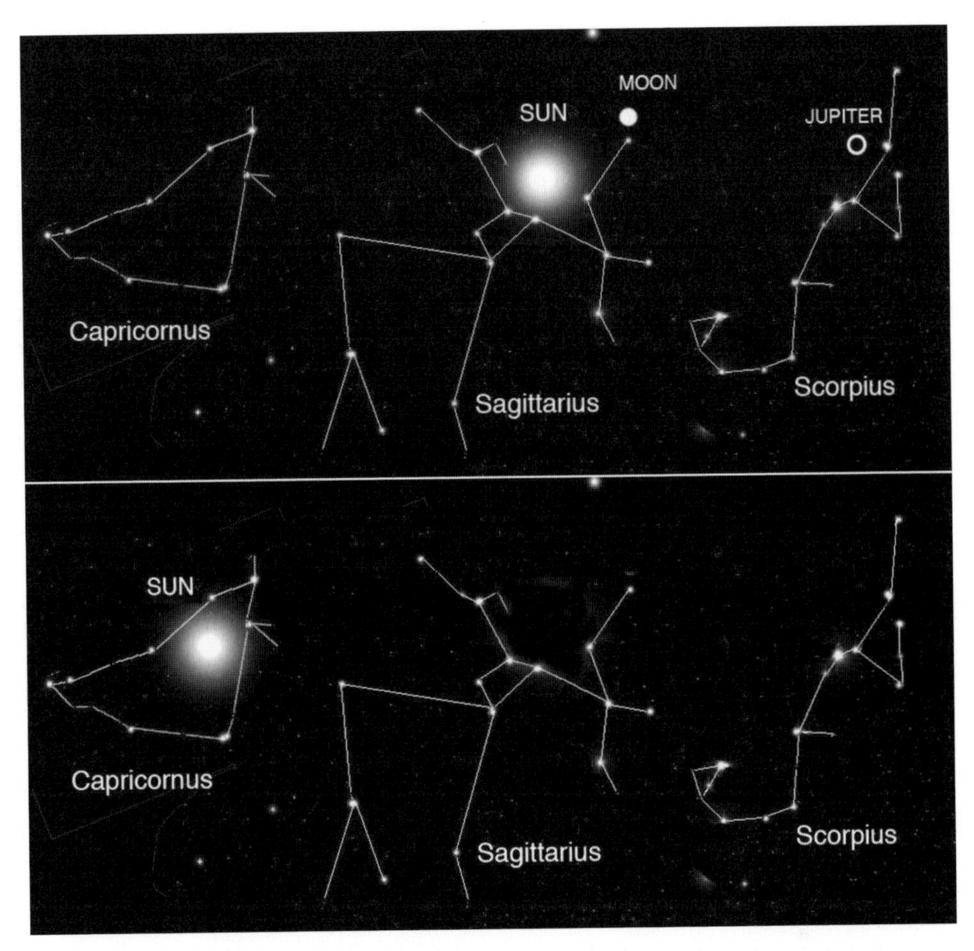

FIGURE 4.15. Maps showing the position of the Sun on 1 January 1995 AD (above) and on 1 January 200 BC, computed using a digital planetarium. According to astrology, on this date the Sun should be in the zodiacal constellation of Capricornus (22 December–20 January). As can be seen by comparing the two maps, this is true only for the sky-map dating back to Hipparchus. In the subsequent 2,200 years, the precession of the equinoxes has moved the Sun backwards, and it is now in the constellation Sagittarius. This simple experiment can be repeated by comparing the position of easily observable planets (such as Jupiter) along the zodiac with astrological predictions. It is also a demonstration of how to use the precession of the equinoxes as a time machine. If the map were to be found during archaeological excavations it would allow a reliable dating of the site.

done by the Gregorian reform, the consequence in the long run would be an uneasy mixing of months and seasons. Precession is therefore confined to the sky, relevant only to a small fraction of the Earth's population: the astronomers. On human timescales, August will always be the warmest month in the northern hemisphere, and spring begins at the end of March, even if over the centuries the

Sun slowly recedes along the zodiacal belt, at a pace of roughly one zodiacal constellation every 2,000 years.

However, the steady growth of the precession leads, in the long term, to macroscopic effects. As an example, the actual position of the Sun is displaced backwards by one zodiacal sign if compared with astrological statements. If those who claim to be born in Virgo were to go into space on their birthday, they would discover that the Sun is in the constellation of Leo. The astrological Leo should be replaced by the Cancer, and so on (Figure 4.15).

Such a striking discrepancy is not completely ignored by astrologers. In his *Practical Treatise of Astrology*, André Barbault, writes: 'No confusion should be made between zodiacal signs and constellations. Once they were overlapping, but due to a phenomenon called precession of the equinoxes the vernal point spans through the twelve zodiacal constellations in a direction opposite to their usual order'. This statement is absolutely correct from an astronomical point of view, but unfortunately it is not followed by a discussion on its consequences for astrology. If the Sun no longer follows the astrological path, then the computation of the position of Sun and planets at someone's birth or at any other time, widely used for astrological predictions, is definitely wrong, and does not represent their true position in the sky. This is why zodiacal 'signs' and 'constellations' are not synonyms; but they do have a rather different meaning. Any hypothesis on the influence of celestial bodies on our lives should account for this unquestionable inconsistency. And the initial question 'What is your zodiacal sign?' should be answered with another question: 'Do you mean my astrological sign or my astronomical sign?' Celestial mechanics makes the difference.

5

Our chaotic Solar System

I do accept chaos, but I'm not sure whether chaos is accepting me.
Bob Dylan

According to a non-scientific dictionary, chaos is 'the original mixture of elements that, following certain cosmogonical theories, existed before the creation of the Universe'. Lucretius – a disciple of the Epicurean doctrine – identifies chaos with the *clinamen* – the small deviation from the parallel free-fall trajectory of atoms in space which makes them encounter each other, create elements and therefore life. After being forgotten and rediscovered many times, chaos is now firmly back in the scientific literature. It plays an important role among apparently distant fields of study: fluid mechanics, medicine, computer science, economy, and quantum systems. The study of chaotic behaviour in celestial mechanics dates back to the pioneering work of Henri Poincaré on the three-body problem, whereas the powerful computing tools at our disposal nowadays allow us to investigate the chaotic behaviour of complex N-body gravitationally interacting systems. In doing so, fast and slow chaotic diffusion emerge as the driving forces shaping the dynamics of our Solar System.

THE UBIQUITY OF CHAOS

The word 'chaotic' summarises many fundamental concepts characterising a dynamical system such as complexity, predictability and stability. But above all it acts as a warning of the difficulties which are likely to arise when trying to obtain a reliable picture of its past or future evolution. As an example, a commonly accepted definition states that a system is 'unstable' if the trajectories of two points that initially are arbitrarily close – thus representing slightly different states of the system – diverge quickly in time. This has strong implications, as small uncertainties in the initial conditions (for example, those deriving from the orbit determination of a celestial body) might be consistent with completely different future trajectories. The conclusion is that we can exactly reproduce the motion of a chaotic system only if we know with absolute precision the initial conditions – a statement that in practise can never be true. This is the way in which chaos hides the 'true' trajectory of a deterministic system.

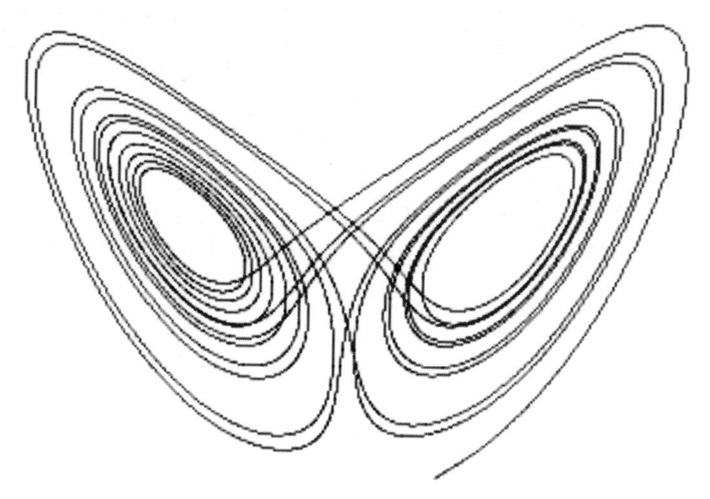

FIGURE 5.1. Example of the Lorenz attractor.

Seagulls and butterflies

In spite of sophisticated software modelling, of the ever-increasing computing power, and of the many meteorological satellites observing the Earth's atmosphere from space, weather forecasting is still subject to a high degree of uncertainty. It is therefore not surprising that this discipline has been responsible for one of the many independent 'discoveries' of chaos.

In 1962 the American meteorologist Edward Lorenz showed that even a relatively simple meteorological system (described by simple mathematics) is subjected to sudden changes from sunny to stormy weather and *vice versa*. In order to illustrate the difficulty of weather forecasting, even in such a simple case because of the extreme sensitivity to the initial conditions, he used the plot shown in Figure 5.1. The transitions from the neighbourhood of one 'attractor' (very good weather) to the other (very bad weather) exhibit chaotic behaviour in the sense that two trajectories, initially very close, would soon diverge, experiencing a completely different sequence of 'jumps' between good and bad weather. In an article presented to a scientific conference Lorenz wrote: 'One flap of a seagull's wing would be enough to alter the course of the weather forever.'

Upon seeing how closely the Lorenz plot resembled a butterfly's wing rather than a seagull's wing, one of Lorenz's lectures, delivered in 1972, was entitled 'Predictability: Does the Flap of a Butterfly's Wings in Brazil set off a Tornado in Texas?' The celebrated 'butterfly effect' was born.

Celestial mechanics has the advantage of dealing with real trajectories in space, and it is somehow easier to visualise the effect of chaos. As we shall discuss in detail later in this chapter and the next, a near-Earth asteroid moves on a chaotic orbit undergoing frequent encounters with the terrestrial planets. Each encounter acts as an amplifier of uncertainty, thus making it difficult to predict

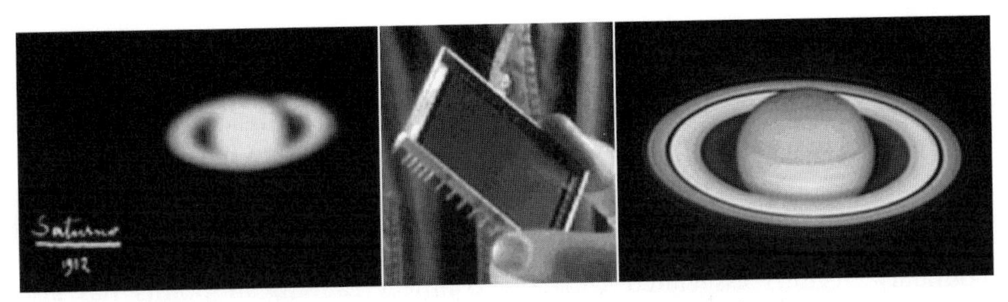

FIGURE 5.2. An example of the advantages that can be achieved by using a CCD (centre): (left) an image of Saturn obtained in 1912 from the Collurania Observatory (Teramo, Italy) (right) an image obtained with a CCD on a modern commercial telescope.

the orbital evolution for timespans longer than a hundred years. On the other hand even a tiny impulse applied to an object on a collision course with our planet and 'amplified' by the very same chaotic nature of the orbit would ensure that it would miss the Earth and not destroy mankind.

Apart from the many probes roaming the interplanetary space and the orbiting observatories looking at the sky at different wavelengths, the last twenty years have also brought significant improvements in the detection of faint objects from the ground. Photographic plates have been replaced by CCDs (charge-coupled devices) – digital arrays so sensitive that they have found a commercial application in modern video cameras (Figure 5.2). The possibility of equipping a telescope with a CCD, and the widespread availability of fast computers and standard image-processing techniques, has allowed not only the professional but also the amateur astronomer to perform high-quality observations. Large telescopes have profited by 'adaptive optics' – a technology that sustains the wide surfaces needed for gathering and concentrating light by splitting a mirror into several smaller components, the orientation being controlled by a computer in order to always maintain the best optical configuration.

The observed number of celestial bodies in the Solar System has increased dramatically: tens of new satellites mostly in 'irregular' high-eccentricity and high-inclination orbits; hundreds of comets with fascinating orbital patterns; tens of thousands of asteroids filling the 'belt' between Mars and Jupiter; puzzling objects such as Chiron and the Centaurs, defying classification as asteroids or comets; the dynamically intriguing population of the near-Earth objects (NEO); and last but not least, a reservoir of comets beyond the orbit of Neptune.

Celestial mechanics has been faced with an unprecedented variety of orbital regimes, often characterised by chaotic evolutionary paths. This is how the origin of meteorites and of the short-period comet population was traced back to the asteroid main belt and to the transneptunian regions, respectively. As we have seen in the previous chapters, chaos has also played a major role in describing the long-term behaviour of the obliquity of the planets, and it is always somehow present, even if to a different extent, any time at which a commensurable motion is involved.

The astronautical sciences have also profited by chaos. Mission designers have learned how to use close encounters with the planets for readdressing the orbital motion of a spacecraft. This technique is called 'gravity assist' because gravitation does most of the work without the need of firing the onboard propulsion and therefore saving consistent amounts of fuel. Riding the 'fuzzy boundary' regions associated with the collinear Lagrangian points has also allowed us to use chaos for tracing novel spaceways to the Moon and the planets.

The ubiquity of chaos has even touched a longstanding problem of celestial mechanics: to determine whether or not the Solar System as a whole is stable. Recent results must once again be discussed in the light of chaos.

Today more than ever it is tempting to join the critic and editor Lord Francis Jeffrey (1773–1850), who once shouted: 'Damn the Solar System! Bad light; planets too distant; pestered with comets; feeble contrivance; could make a better one myself!'

PROPAGATING ORBITS

The observed orbits of the planets are distant enough to avoid collision. But for how long has this been so? And more importantly, are we sure that this scenario will not change in the future? The reason why it is not easy to answer these questions is that in order to compute the orbital evolution of each individual planet, all relevant perturbations must be taken into account. This results in the full *N-body problem*, with equations of motion that are non-integrable and difficult to handle from a mathematical point of view.

After having resisted the most skilled astronomers of the past, the quest for the 'Holy Grail' of celestial mechanics gained momentum after the development of modern high-speed digital computers. For the first time it has become possible, in principle, to compute the dynamical evolution of all planets for a timespan comparable to the age of the Solar System: five billion years. The excitement at having reached such a distant frontier has led to ambitious research projects which require the construction of computers designed to perform only one single very specific task: to move planets along their orbits. They are called 'digital orreries', analogous to mechanical orreries which made model planets move by means of complex devices. But moving planets 'electronically' along their orbits for long timespans produces huge amounts of data which must be archived, requiring the development of specific software programs for retrieval and analysis. In facing this problem it was soon realised that it is useless to let a computer run free without questioning the significance of the data obtained, as the complex interplay of the accuracy with which perturbations are modelled, and the level of chaos which characterises the orbits, could cast a shadow on the results.

As we have seen in the case of main-belt asteroids, apparently stable orbits become strongly chaotic on timescales of the order of tens or hundreds of millions of years, thus compromising the reliability of the results for longer timespans. But in the long-term even a much lower level of chaos introduces a large uncertainty. A weakly chaotic planetary orbit will never undergo dramatic changes of the orbital parameters, thus ruling out any risk of catastrophic collision with the neighbouring planets. In this case the orbit is so sensitive to perturbations that no long-term predictions of the position along the orbit can be made.

This is why the methods developed by celestial mechanics for orbit propagation range from sophisticated mathematical theories (time series expansions, Lagrange planetary equations, and so on) to step-by-step procedures that foresee the extensive use of computers (numerical integrations). The former approach is called *general perturbation methods*, while the latter are referred to as *special perturbation methods*. Each method has its own advantages and limitations. General perturbation methods are in the form of long expressions which can be analysed term by term, thus showing more clearly perturbations at work and deriving precious information on the dynamical characteristics of the system. However, such methods are usually implemented on model problems, thus introducing some approximations with consequences that are difficult to evaluate in advance. The dynamical system under study is therefore not the 'real' one, but should be as close to it as we like. The problem is that it is not always possible to demonstrate that it actually remains as 'close as we like' for all cases!

Special perturbation methods apply the numerical solving of ordinary differential equations to those describing gravitationally interacting systems. This procedure implies no mathematical handling of the equations of motion, but it can be very slow due to the large number of time steps involved. Furthermore it introduces round-off and truncation errors, as a consequence of the finite number of significant digits used by the computer and of the intrinsic accuracy of the method.

More on numerical integrators

Among the most common integration methods used in celestial mechanics it is also worth mentioning *extrapolation methods* and *symplectic integrators*.

Extrapolation methods provide an estimate of the magnitude of the error during the integration, and use this information to improve their own accuracy. In doing so they allow a substantial increase in the step size by controlling the propagation error, thus increasing speed in most cases.

Symplectic integrators are increasingly used for long-term numerical integrations, because they are able to preserve some key features of the N-body problem. In particular, theoretical considerations require that the energy of any gravitational system remains constant in time. In general the approximations introduced by the numerical integrations do not guarantee the conservation of energy, while symplectic integrators are especially designed to this end. This is why they are also called *energy preserving* methods.

A simple test can be performed to evaluate the errors introduced by numerical integration. After computing the forward evolution of a system the procedure is reversed, and the integration is repeated backwards for an identical timespan. In general, going numerically 'back and forth' does not result in a return to the beginning. The difference is called *closure error*, which provides an estimate of the accuracy of the method adopted.

The techniques described so far have been successfully applied to real cases involving celestial bodies, artificial satellites and spaceprobes. The choice of integration method essentially depends on both: the dynamical characteristics of the system, and the nature of the problem under investigation. The sophisticated lunar theories developed at the beginning of the twentieth century and, more recently, the resonance-based model responsible for the confinement of Neptune's ring arcs, are known examples of the achievements obtained by means of general perturbation methods. The ephemeris numerically generated at the Jet Propulsion Laboratory, in Pasadena, using extremely accurate special perturbation methods, are state-of-the-art in the computation of the precise positions of all major Solar System bodies over several centuries.

PLANETS IN NUMBERS

The geocentric system – in which the five planets known in ancient times, plus the Sun and the Moon, revolved along Earth-centred circular orbits – can be considered a first attempt to grant stability to the Solar System as a whole. The Copernican revolution and the new pillars of celestial mechanics – Keplerian motion and gravitation – strengthened this view, because they could demonstrate – not simply believe – that planets move on orbits that do not intersect and are so weakly perturbed that they can be safely approximated by ellipses with small eccentricity. More importantly, the direct relationship between the period of revolution and the semimajor axis of a celestial body, expressed by Kepler's third law, could be used to correctly estimate the mean distance of the planets from the Sun.

The astronomers of the mid-eighteenth-century were the first to be faced with data on the true distances of the planets (see Table 5.1). At that time, 'playing' with numbers was rather common among scientists, and it was soon realised that planetary distances could be reproduced by the following steps:

- Take the series of numbers 0, 3, 6, 12, 24, 48, 96.
- Add 4 to each number of the series.
- Divide each result by 10.

As can be checked with Table 5.1, the numbers obtained in this way (0.4, 0.7, 1.0, 1.6, 2.8, 5.2 and 10.0) are remarkably close to the actual distances of the planets – taking care to discard a term when passing from Mars to Jupiter. The series listed at point 1 is well known to mathematicians as a 'geometrical progression', and every term but the first is obtained by multiplying by 2 the preceding one.

Possibly because of this, the coincidence was considered only a curiosity until Johann Daniel Tietz (1729–1796) (in latinised form, Titius) and Johann Elert Bode (1747–1826) began to wonder whether there was any underlying dynamical meaning.

Table 5.1. The Titius–Bode law compared to the true astronomical distances.

	Titius–Bode law (AU)	Distance (AU)
Mercury	0.4	0.39
Venus	0.7	0.72
Earth	1.0	1.00
Mars	1.6	1.52
Asteroids*	2.8	2.77
Jupiter	5.2	5.20
Saturn	10.0	9.54
Uranus*	19.6	19.19
Neptune*	38.8	30.07
Pluto*	77.2	39.48

*Not known at the time of Titius and Bode

The law of planetary distances

A compact formulation of the law of planetary distances is obtained as an algebraic rule based on powers of 2. Denoting by d_n the distance of the nth planet, and using the values $n = -\infty$ for Mercury, $n = 0$ for Venus, $n = 1$ for the Earth, and so on, the resulting expression is

$$d_n = 0.4 + 0.3 \times 2^n,$$

recalling that $2^{-\infty} = 0$, $2^0 = 1$, $2^1 = 1$, $2^2 = 2 \times 2 = 4$, $2^3 = 2 \times 2 \times 2 = 8$, and so forth. This simple rule is known as the Titius–Bode law, or simply as Bode's law.

The naming of this rule has an intriguing story behind it. In 1766 Titius was completing the translation of the book *Contemplation of the Nature* by the French natural philosopher Charles Bonnet. At that time the whole process of publishing and translating a book took years of work, and in the meantime new results could possibly appear. It was therefore customary for the translator to add notes to the original text, and sometimes changes were introduced without referring to the author nor the reader. Titius was not alone in knowing the strange relationship among the planetary distances, but he was so enthusiastic about it that he added a note to the book that he was translating. After publication, a copy of the book reached Hamburg, and the note written by Titius caught the attention of Bode. He was so interested that he began publishing articles on it, without referring to his source. Titius protested, but his position was weak, as the law of planetary distances was not an original idea of his own, and he had simply made it more widely known. Eventually, as often happens in the scientific community, both contributions were acknowledged.

In 1781 William Herschel, together with his sister Caroline, found the seventh planet of the Solar System, Uranus. The value of the mean distance from the Sun of the new planet (19.19 AU) was in good agreement with the law of planetary distances. The additional term corresponding to $n = 6$ was, moreover, readily computed as $d_6 = 0.4 + 0.3 \times 2^6 = 19.6$. This result strengthened the belief that the law was not a mere coincidence; and in particular Bode remarked that the jump from $n = 2$ to $n = 4$ when passing from Mars to Jupiter was a clear indication of the existence of another celestial body at a distance corresponding to $n = 3$ (2.8 AU). The search for the 'missing planet' began with great expectations. After all, it had been possible to detect a much more distant planet such as Uranus.

Over the following two decades the awaited discovery did not occur – until in the early hours of the night on New Year's day 1801 the Italian astronomer Giuseppe Piazzi found Ceres from the Palermo Observatory (Sicily). Orbit determination showed that the new celestial body was at the correct distance to fill the gap in the Titius–Bode law, yet its luminosity was too low for it to be of planetary size. Soon, other celestial bodies were found (Pallas, Juno and Vesta in 1802, 1804 and 1807 respectively) sharing the same region of space, and it became clear that they were 'unusual' planets. William Herschel named them 'asteroids'. The problem of the 'missing planet' then turned to finding a plausible origin of the asteroid population. Were they the result of a cosmic catastrophe which had completely disrupted the once fifth planet from the Sun? Whatever the answer, the spectacular confirmation of the validity of Bode's law led to astronomers' confidence that other planets awaited discovery at the distances obtained by prolonging the geometrical series behind the law.

Neptune was discovered in 1846, and celestial mechanics played a leading role in predicting its position in the sky. Yet it represented the first failure of the law of planetary distances, as the new planet should have been orbiting at 38.8 AU, while in reality Neptune's distance from the Sun is only 30.07 AU. (This disagreement justifies the mistake of Adams and Leverrier as discussed in Chapter 1, since they assumed that the 'Uranus perturber' had a semimajor axis of 38.8 AU. Such a large deviation from Bode's law was totally unexpected at the time.)

Almost a century later, in 1930, the American astronomer Clyde Tombaugh discovered Pluto, and the discrepancy grew even larger: 77.2 AU according to the law of planetary distances, against about 40 AU as derived from observations.

Today the Titius–Bode relationship has lost much of its fascinating mystery. The attempts to uncover a physical meaning hidden behind its numbers have been frustrated over the years. And yet modern celestial mechanics does not consider it a mere curiosity, as numerical modelling of the formation of planetary systems has shown that relationships like that introduced by Titius and Bode are common – but the explanation is possibly not as simple as the algebraic relationships describing them.

Piazzi's serendipity

The discovery of the first asteroid can be considered as a perfect mix of serendipity and professional skills. Piazzi was not searching for the missing planet. For ten years he had been dedicating himself to compiling a stellar catalogue down to 8th magnitude, with the aim of its becoming the most complete source to date. In 1787, after returning from a long trip to France and England, he took with him an excellent telescope built by Jesse Ramsden in London, after which he devoted all his time and energy to mapping the sky – which is why he was observing on New Year's Day. In order to produce accurate positions and to minimise errors, he did not observe a stellar field only once, but repeated the same measurements during at least two subsequent nights. He was therefore able to recognise that a faint star observed in Taurus on 1 January appeared in a slightly different position on the following night. Thinking that the early observation was affected by an error, he observed it again on 3 January; and it had moved in the same direction and at the same rate as before. It is then reported that at this precise moment Father Gioacchino Giuseppe Maria Ubaldo Nicolò Piazzi, Abbot of the Teatini order and Professor of Astronomy and Director of the Royal Astronomical Observatory of Palermo, began jumping on the terrace of the observatory and shouting 'A discovery! A discovery!'

THE STABILITY OF THE SOLAR SYSTEM

From this short historical review it emerges that the stability of the Solar System is the result of the interplay of many complex phenomena, more entangled than it was previously thought. In particular the late stages of planetary accretion are largely responsible for the present dynamical settlement. When Earth-sized protoplanets appeared on the scene the interplanetary space was still filled with planetesimals – chunks of irregularly shaped rock a few kilometres wide. Inside such a crowded Solar System, orbits are unstable, and the probability of encounter is high, leading eventually to impacts, catastrophic collisions or ejections into interstellar space. Exotic dynamical effects, such as migration of the orbits of the planets, occur as the result of these strong and/or frequent gravitational interactions. The subsequent evolution leads to a more stable situation in which matter is concentrated in a few massive bodies – the planets – travelling through a relatively empty space. The battered surfaces covered by impact craters on the Moon, Mercury, the asteroids and many satellites show the scars of a long collisional history (Figure 5.3). Moreover, a criterion widely used to evaluate the age of a celestial body is to analyse the number and the size of the impact craters present on its surface. From this simple crater count the flux of impactors can be traced from the heavy bombardments characterising the early Solar System to the present sporadic cratering events.

The effect of purely gravitational perturbations acting throughout the ages is

FIGURE 5.3. Impact craters reveal the evolution of the Solar System. Saturn's satellite Mimas (left) bears the sign of a large impact, just below the threshold for complete fragmentation. (Courtesy NASA/JPL) Even a small region of the lunar surface (centre) shows an impressive variety of cratering events, from large, old craters with smoothed edges to circular bowl-shaped craters formed by recent meteoritic impacts. The flat muddy bottom of martian craters (right) provides evidence of a lake that once filled the depression. (Courtesy ESA.)

also difficult to reconstruct, as is the fate of the Solar System. Despite the achievements of celestial mechanics over more than two centuries and the computing facilities available nowadays, the stability of the Solar System is still open to question.

A first-order problem is to agree on what 'stability' actually means. It could be satisfied with a sufficiently long regular motion, while conversely a rigorous proof of stability valid over an infinite time would be highly desirable. The difference in dealing with a limited period of time, however long, or with infinity, is in deciding both from a mathematical and a physical point of view.

In Chapter 2 it is pointed out that KAM theory guarantees the stability of a dynamical system over an infinite time at the price of being subjected to strict limitations on its applicability to real systems. Numerical investigations are constrained by the large amount of computer time involved and by an intrinsic lack of accuracy. A good compromise could be to restrict it to the estimated lifetime of the Solar System. From astrophysics we know that the Sun is presently in the middle of its life, and that 5 billion years from now it will evolve into a red giant, growing in size until encompassing the orbit of Venus before ending as a white dwarf. A significant result for celestial mechanics could therefore be to establish the past and future stability of the Solar System over a timespan of the order of 10 billion years.

In 1988 a technological challenge was undertaken by Gerald Sussman and Jack Wisdom at the Massachusetts Institute of Technology, in Boston. Their project was to build a computer that had hardware especially designed for carrying out simulations of the long-term dynamics of the outer solar system, from Jupiter to Pluto. Their digital orrery carried out an integration of the corresponding N-body equations of motion over a timespan of 845 million years. It showed that the orbits of Jupiter, Saturn, Uranus and Neptune are rather stable, but that Pluto

Stability after Laplace

Towards the beginning of the nineteenth century the mechanical interpretation of the Universe was in full development. Any physical phenomenon – from falling bodies to heat transport, and from electricity to light – was ultimately explained as the mechanical interaction among particles. In this respect the French mathematician Pierre Simon de Laplace was by no means an exception. He was so confident in celestial mechanics to approach the problem of the stability of the Solar System by means of perturbation theories. His studies led him to claim that an overall dynamical invariance of the Solar System could be found, relying on the fact that the mutual gravitational interactions among the planets were modifying, only to a minor extent, their trajectories. Unfortunately modern computer simulations showed that the computations performed by Laplace were not sufficiently accurate to support his alleged stability result. Nevertheless, his pioneering investigations into the stability of the Solar System represented a milestone in the subsequent development of perturbation techniques. Furthermore, his work inspired Adams and Leverrier in their studies which eventually led to the discovery of Neptune.

exhibits chaotic behaviour, and as a consequence its motion cannot be reliably predicted over long timescales.

At the same time, Jacques Laskar, at the Bureau des Longitudes in Paris, implemented a fast computer to run an advanced perturbation theory that allowed him to integrate the motion of all the known planets (except Pluto) over a period of 200 million years. The results confirmed an overall regularity of the motion of the giant planets from Jupiter to Neptune. On the other hand the behaviour of the inner planets – Mercury, Venus, Earth and Mars – appeared to be definitely chaotic. In particular, if for any one of them two separate numerical integrations are started from very close initial conditions, the distance between the trajectories grows by a factor 3 every 5 million years. This does not necessarily imply a future scenario of catastrophic collisions among the planets, but it must be admitted that these conditions prevent a precise positioning of the inner planets over a long timespan. Although the orbital path of our planet will never experience major changes, it is not possible for us to compute where the Earth will be along its orbit 100 million years from now.

The ability to distinguish different degrees of 'chaoticity' is becoming increasingly important for celestial mechanics. 'Soft chaos' – such as that characterising the motion of the inner planets – is obviously a completely different drawback on the stability of a system. To this end, measurement of how quickly two initially nearby trajectories are diverging in time is a useful parameter. The 'Lyapunov indicators' (after the Russian mathematician Aleksandr Mikhailovich Lyapunov, 1857–1918) satisfy this need, and are now an indispensable tool when investigating the complex relationship between stability and chaos.

CLOSE ENCOUNTERS

Comets were the first objects to show clear chaotic behaviour. As soon as it became possible to provide an insight into their orbital evolution, peculiar regimes of motion were found, characterised by frequent close encounters with the planets and consequent abrupt changes of trajectory.

This might be surprising, as the regular apparitions of comet Halley – possibly the most widely known of the comets – have marked the development of our civilisation (Figure 5.4). It is named after the second Astronomer Royal Edmond Halley (1656–1742), who showed that comets can travel on elliptic orbits and that the apparitions of 1531, 1607 and 1682 were different returns of the same object. The first accurate prediction of its position in the sky was obtained by Philip Herbert Cowell (1870–1949) – one of the pioneers of the numerical integration of the orbits – at the 1910 apparition. It represented a major advance in celestial mechanics, because it demonstrated the validity of perturbation methods in a practical case. Yet comet Halley is somehow an exception for the typical behaviour of comets, and a more representative example is the dynamical history of the much less famous comet Lexell.

On 14 June 1770 the celebrated French astronomer Charles Messier (1730–1817), while observing from the Naval Observatory of Cluny, in Paris, added a new entry to the long list of comets that he had discovered. It had a period of revolution around the Sun of about 5.5 years, but the comet was never to be seen again. What happened?

An explanation was provided by the Swedish astronomer Anders Johan Lexell (1740–1784), who investigated the motion of the comet during the years immediately preceding and following its discovery. His work is a striking example of the incredible amount of information that can be provided by celestial mechanics. First of all, Lexell pointed out that the lost comet had a very

FIGURE 5.4. Edmond Halley, and his comet in an illustration dating from 1066.

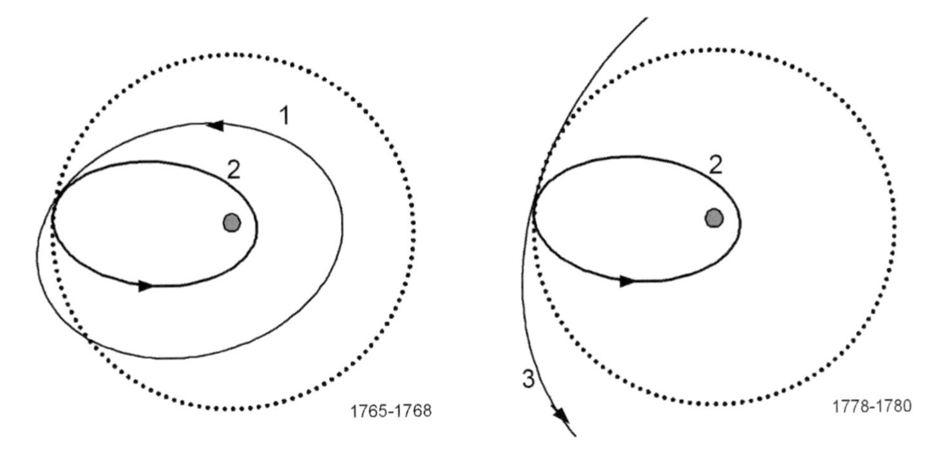

FIGURE 5.5. Comet Lexell's two close encounters with Jupiter (orbit indicated by a dotted circle). The former (left) reduced the size of the orbit (when passing from orbit 1 to orbit 2), thus allowing its discovery, while the latter resulted in the ejection of the comet from the inner Solar System (orbit 3).

short period of revolution, and that its aphelion was not too far from Jupiter's orbit, thus allowing encounters with that planet. When computing the orbital evolution Lexell found that in the past the comet had travelled on a larger orbit, with a perihelion near 3 AU (well beyond the orbit of Mars), thus being too faint for Earth-based detection. In 1767, just before the discovery, the comet passed very close to Jupiter, which caused a major perturbation of the comet's trajectory. The orbit shrank, and the perihelion distance was reduced to less than 1 AU. Due to the comet's small distance from Earth, together with the onset of cometary activity and the consequent formation of a tail, it was easily observable in 1770.

Some years later, just before arriving at aphelion, the comet again encountered Jupiter, due to the synchronising effect of a 1:2 mean motion resonance. This encounter with the giant planet was so close that the comet crossed the orbits of the Galileian satellites. Lexell exploited this event for deducing that the mass of the comet must have been very small, since the motion of the satellites did not exhibit any perturbations. This beautiful reasoning can be considered the first estimation of the size of a comet nucleus.

On this occasion the strong perturbations exerted by Jupiter resulted in placing the comet on a much larger orbit (Figure 5.5) extending beyond the farthest planet and therefore requiring hundreds of years to be completed. This is why a return had been waited in vain. Goodbye comet Lexell...

We know that close encounters with the planets are rather frequent and rule the orbital evolution of comets, representing the key for understanding their origin. Far back in the early era of planetary formation, the outer Solar System was crowded with 'cosmic icebergs': the icy planetesimals. Those not contributing directly to planetary accretion were undergoing close encounters with the

What's in a name?

Naming comets is a tough business. With a few notable exceptions (Halley and Lexell are named after the astronomers who studied their orbits), a comet takes the name of the person who has discovered it – a simple rule with complex consequences. A systematic search for new comets involves long nights spent sweeping the same region of the sky, waiting for a small intruder to appear, and it is therefore not surprising that there are simultaneous discoveries of the same object by two observers. When this happens, both names are given to the comet (a recent example is that of comet Hale–Bopp). If a comet is lost for some time due to bad observing conditions or due to a lack of cometary activity, the name of the 'rediscoverer' is also added to the list. As a consequence, a comet's name is often a long sequence of surnames of different nationalities, such as Honda–Mrkos–Pajdusakova, Tuttle–Giacobini–Kresak, DuToit–Neujmin–Delporte, and West–Kohoutek–Ikemura (Figure 5.6). Moreover, when the same discoverer finds more than one comet a sequential number is added. Language-breaking exercises are the comets Tsuchinshan 1 and Tsuchinshan 2, and the stiff case of Schwassmann–Wachmann 1, Schwassmann–Wachmann 2 and Schwassmann–Wachmann 3, originated by the successful collaboration between the German astronomers Arnold Schwassmann and Arno Harold Wachmann.

These general rules have been recently revised due to the increasing number of comets found by automated telescopic surveys, such as the US project LINEAR (Lincoln Near Earth Asteroid Survey) or by orbiting telescopes (the European solar mission SOHO has already discovered more than 1,000 comets). Comets are now numbered sequentially, while additional letters indicate their 'type': P/ for periodic comets, C/ for non-periodic, D/ for objects which no longer exist, and X/ for those with orbits not yet determined. For example, 82P/Gehrels 3 means that the third comet discovered by the American astronomer Tom Gehrels belongs to the short-period comet population and bears the catalogue number 82.

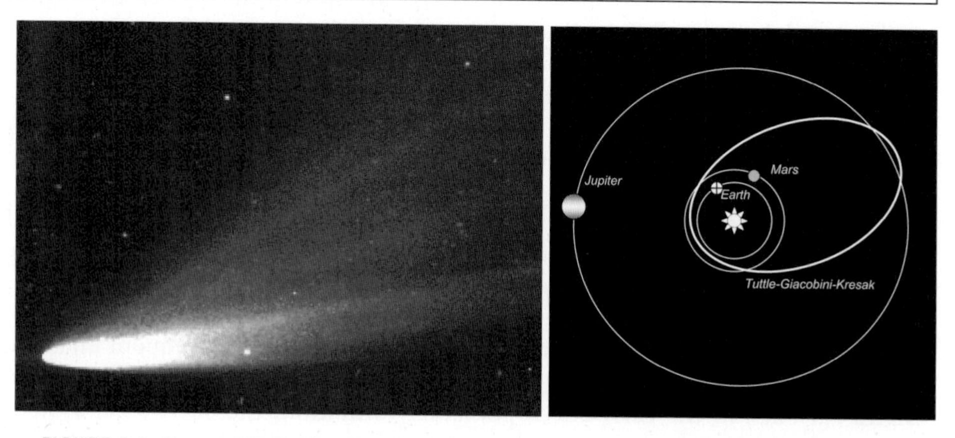

FIGURE 5.6. Comet 41P/Tuttle–Giacobini–Kresak belongs to the Jupiter family of comets, as indicated by its orbit, the aphelion of which is almost tangential to Jupiter's orbit.

giant planets. This process resulted in the surrounding of our planetary system with the 'building blocks' of planets discarded during their construction – a distant repository located at about 50,000 AU from the Sun (a fifth of the distance to the closest star). This hypothesis was proposed by the Dutch astronomer Jan Oort (1900–1992) in the early 1950s, and is known as the Oort Cloud.

At such great distances from the Sun, each member of the cloud is very sensitive to the gravitational perturbations of passing stars or giant molecular clouds, as well as to the effect of the galactic potential, which may induce orbital changes strong enough to bring it back into the planetary region on highly eccentric orbits. If the comet reaches a distance from the Sun roughly comparable to that of Mars, the ice, heated by the Sun, starts outgassing violently. The dust and particles present in the nucleus are also ejected into space and, reflecting the sunlight, produce the spectacular coma and tail. The tiny, ugly planetesimal transforms into a beautiful comet extending over millions of kilometres.

RAINING COMETS

Oort's model accounts for only a fraction of the whole population of comets; namely those on long-period, high-eccentricity and inclined orbits. Many other comets have moderate eccentricities and short periods of revolution – thus the term *short-period comets.* In order to explain their existence, the so-called 'stepwise capture' seems attractive. Travelling towards the inner Solar System, an Oort Cloud comet can undergo a sequence of planetary close encounters. For example, Neptune can modify a comet's orbit in such a way that the comet meets Uranus, which in turn redirects the comet to Saturn, after which it eventually moves into Jupiter's gravitational domain. As seen in the case of Lexell's comet, Jupiter is very efficient in reducing the perihelion distance of a comet so that it is easily observable from Earth.

Unfortunately this fascinating scenario – a true cosmic game of billiards – is not fully consistent with the observed flux of short-period comets. Indeed, in the second half of the twentieth century the Uruguayan astronomer Julio Fernandez pointed out that the 'stepwise capture' was not efficient enough to justify the number of comets presently observed in short-period orbits. The 'rain' falling from the Oort Cloud was not consistent with observations.

This result provided the dynamical ground to another hypothesis: the existence of an additional reservoir of comets located just beyond Neptune, as proposed independently by Gerard Kuiper (1905–1973) and Kenneth Edgeworth (1880–1972). The observational evidence was eventually found in 1992 with the discovery of the first transneptunian object (TNO). The number of known TNOs is presently approaching 1,000, thus confirming that a whole new population of icy celestial bodies occupies the extreme regions of the Solar System. Their dynamics is characterised by resonances, chaotic behaviour and collisional

FIGURE 5.7. The nucleus of comet Halley, as imaged by the Giotto spacecraft. For the Halley fly-by event in 1986, both Fred Whipple (top right) and Jan Oort were special guests at the Giotto mission control centre. (Courtesy ESA.)

evolution, and their relevance for celestial mechanics extends far beyond providing the required parent bodies for the short-period comets. Many TNOs have orbital parameters closely resembling Pluto's, and some have comparable if not larger sizes. In a following chapter we will discuss in detail these new members of our Solar System.

The spectacular images of comet Halley provided in 1986 by the European Giotto spacecraft, and the many subsequent missions to comets allowing close-range imaging, have proven that the nuclei of short-period comets are indeed small kilometre-sized icy bodies. These results, while confirming the 'dirty snowball' model proposed by the American astronomer Fred Whipple (1906–2004), added a new chapter to the cometary saga (Figure 5.7). From a physical point of view the icy nucleus of a comet is periodically subjected to extreme environments: the cracking of the surface induced by solar radiation triggers violent jet streams, and the close encounters with the giant planets induce strong internal tidal stress. This explains why comets have been often observed splitting into two or more components, and sometimes completely disintegrating.

A well documented historical case is comet Biela, which separated into two components in 1846, reappeared as a double object in 1852, and was never seen again. Instead, an intense meteor shower occurred at what would have been the

1872 return – thus indicating that the comet had disintegrated and confirming the link between comets and meteors.

In more recent times, the spectacular break-up of comet Shoemaker–Levy 9 into more than twenty major fragments, and their impact on the atmosphere of Jupiter in July 1994, was followed with world-wide live coverage. Images and movies captured by the large telescopes on the ground and by the Hubble Space Telescope quickly spread through the Internet, allowing a wider audience to witness the cosmic event.

The relevance for celestial mechanics is that when escaping such a dramatic fate as that of comet Shoemaker–Levy 9, some short-period comets could well be fragments of the same parent comet. However, the different fragments will keep following similar orbital patterns for only a relatively short time, as chaos makes them quickly spread, thus hiding any common origin.

An illuminating example is that of the two periodic comets Neujmin 3 and Van Biesbroek, independently discovered, respectively, on 2 August 1929 from the Observatory of Simeiz (Crimea), and on 1 September 1954 from the Yerkes Observatory (Wisconsin). When the orbital motions of the two comets were integrated numerically, it was found that their past dynamical history was surprisingly similar before a close encounter with Jupiter occurred in 1850. The similarity was so striking that it could not be due to chance, and the obvious explanation is that the two comets are, in reality, the fragments of a larger comet disrupted by tidal forces during the encounter with Jupiter.

The study of the origin of comets dates from the end of the sixteenth century, when Tycho Brahe was the first to demonstrate that comets were not meteorological phenomena, but rather were celestial bodies travelling through interplanetary space. Yet the analogy with terrestrial weather has somehow remained associated with comets as we speak of dirty snowballs, the Oort Cloud and meteor showers.

THE LONG JOURNEY OF METEORITES

The exploration of the Solar System with automated spacecraft has succeeded in remotely sensing the surfaces and the atmospheres of the celestial bodies, and in many cases, landing upon them. A major step ahead is to bring back samples of alien worlds to Earth. The collection of lunar rocks and bringing them to Earth represented one of the major achievements of the Apollo missions, because it allowed scientists to analyse them in well-equipped laboratories. A Mars sample return mission is presently under study, but it must face the complex problem presented by the need of protecting our planet from possible biological contamination. In this context, the Stardust mission has succeeded in collecting the grains of a comet's tail in a small capsule, which, in a sealed container, was safely brought back to Earth on 15 January 2006 (Figure 5.8). Before the space age, meteorites were the only samples of extraterrestrial material available on our planet.

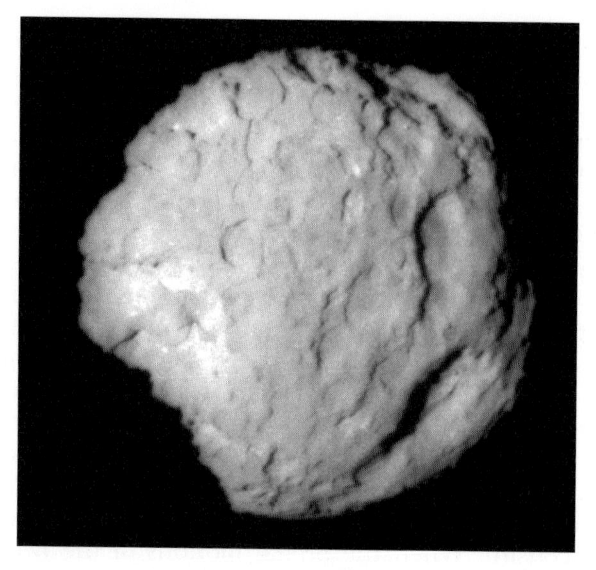

FIGURE 5.8. The nucleus of the Stardust mission target, periodic comet Wild 2, closely resembles a 'dirty snowball'. (Courtesy NASA.)

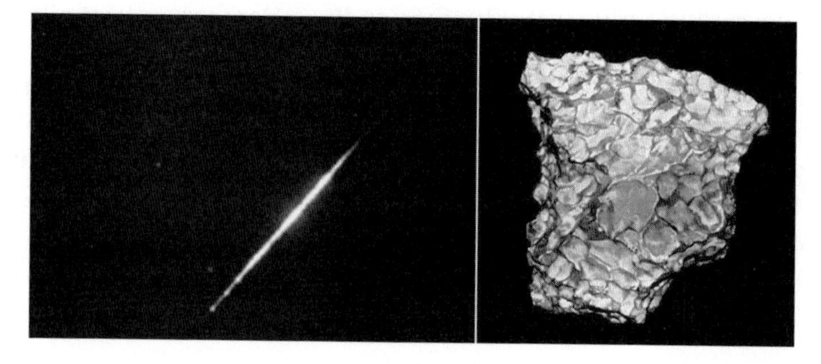

FIGURE 5.9. The origin of meteorites. The fragments produced during a major impact in the asteroid main belt undergo chaotic diffusion toward the inner Solar System, eventually entering the Earth's atmosphere and leaving long bright trails (left). If the fragments are large enough to survive the heat generated by friction with the air, they can reach the ground (right).

A meteorite is the remnant of a wandering celestial object large enough to have survived the heat generated by high-velocity entry in the Earth's atmosphere, and small enough to have reached the ground without causing significant damage (Figure 5.9). The fall of a meteorite is often witnessed by the long bright trail left in the sky, often accompanied by loud bangs. Sometimes the pieces can be found and recovered, and meteorite collections have been instituted as a means to preserve and study our only confirmed 'visitors' from outer space.

Meteorites are traditionally classified according to their chemical composition and their microscopic structure: stony meteorites are rocky aggregates, metallic meteorites are rich with iron and nickel compounds, and others seem to have undergone melting processes and are similar to the lava erupted by volcanoes. To a geologist these simple differences reflect the internal structure of a planet: a rocky crust, an underlying molten mantle and a metallic core. To an astronomer this analogy is a reminder of the hypothesis on the origin of the multitude of irregularly shaped asteroids dispersed between the orbits of Mars and Jupiter. A straightforward conclusion could be that meteorites are the smallest by-products of a cosmic catastrophe: the break-up of a planet massive enough to have undergone large-scale differentiation, responsible for the formation of the asteroid belt. Rocky, basaltic and metallic meteorites originate respectively from the surface, the mantle and the core of the disrupted planet. Moreover, when ground-based spectroscopic observations were used to analyse the light reflected from the asteroids, their bulk composition proved a good match with the compositional types of the meteorites. Nevertheless, two fundamental questions remained unexplained. How could an entire planet break into pieces? And by which dynamical process could a small fragment leave the asteroid belt and reach the Earth?

The lack of a reasonable answer to the former question has resulted in the development of less catastrophic theories on the formation of the asteroid belt. It is now believed that early in the history of the Solar System the planet Jupiter, perturbing, with its huge mass, the region where the asteroid belt is now located, increased the relative velocity among planetesimals, thus inhibiting the growth of a large planet. The result is a collisional system in which disruptive encounters among its members took place in the past and are still happening today. Some of the larger bodies, such as Ceres or Vesta, have escaped complete fragmentation and are almost spherical, although on the latter a large crater with a central peak has recently been identified. However, the vast majority of asteroids are irregularly shaped bodies ranging in size from a hundred kilometres to boulder-sized debris.

The traces left by these collisional events in the distribution of the asteroids are represented by distinct groups of objects having similar orbital elements. They are called *asteroid families* (Figure 5.10), as each group was possibly formed by fragments with different size and shape originating during a catastrophic impact. In this respect meteorites are just the smallest members of a family, landed on Earth after a long journey through interplanetary space.

This scenario has a common origin with asteroids and meteorites, and at the same time provides an overall convincing picture of what happened in the region of space between the orbits of Mars and Jupiter. Unfortunately it is quite difficult to identify which celestial route the asteroidal fragments should follow in order to reach our planet and enter the atmosphere. The ejection velocity of a rocky fragment generated during an impact is typically not sufficient to insert it into an express trajectory to Earth, and so more complex dynamics must be at work.

The long-sought answer was eventually found through the combined action of resonances, chaos and a subtle dynamical mechanism known as the *Yarkowsky*

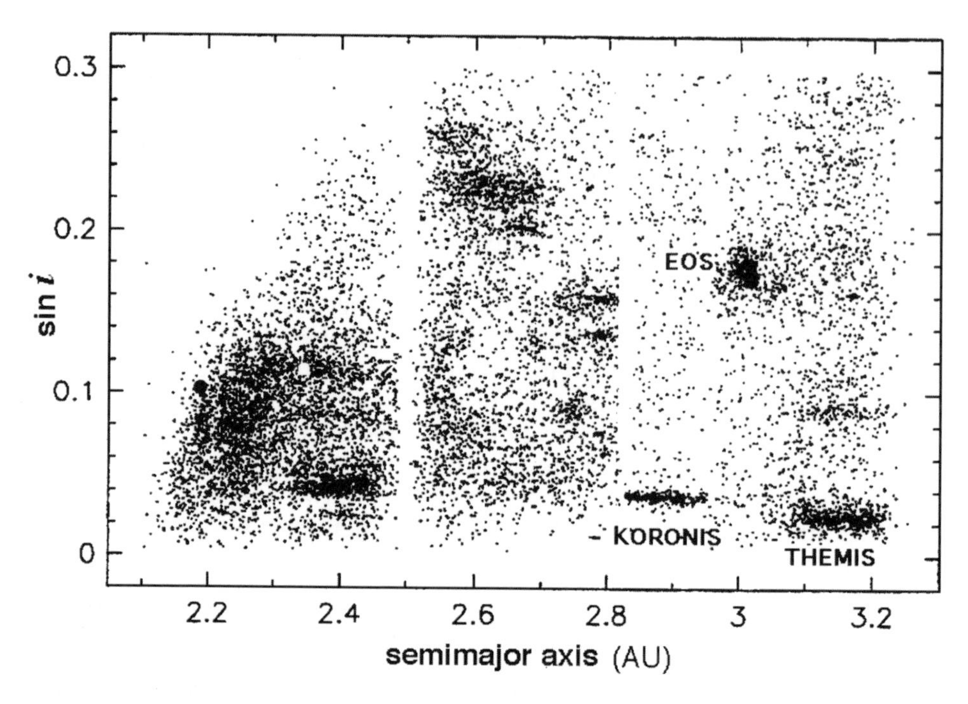

FIGURE 5.10. The distribution of asteroids is not uniform, but exhibits distinct clustering, interpreted as collisional families. (The inclination is plotted as a function of mean distance from the Sun). The three largest families – Themis, Eos and Koronis – were identified for the first time by the Japanese astronomer Kiyotsugu Hirayama (1874–1943). They are named after the largest asteroid or the parent body of the family. The long empty stripes show the location of the Kirkwood gaps.

effect. The whole process is similar to the formation of the Kirkwood gaps, whereby the fragments generated by an impact can be injected inside a resonance and can exploit chaos to began their travels. Chaotic evolution, including planetary close encounters, cause major changes of their orbits until a collision trajectory with the Earth is achieved.

The problem with this explanation is time. The 'age' of a meteorite - how long it has been wandering in space - can be measured through changes in its composition due to exposure to energetic cosmic rays. For most meteorites, laboratory data indicate exposure times of the order of 100 million years, while the chaotic highways of celestial mechanics are run at least tens times faster. The discrepancy was settled only recently, due to the rediscovery of a tiny force acting on small celestial bodies, which accumulates over long timespans producing significant dynamical effects: the Yarkowsky effect, named after the Russian engineer who first proposed it at the beginning of the twentieth century. It is related to the difference of temperature between the warmer 'afternoon' side of a rotating celestial body with respect to the colder 'morning' side. The consequence is an asymmetry in the heat emission, which produces a force so small that it should be taken into account only for celestial bodies with diameters less than 10 km over long timespans. This was exactly what astronomers were looking for to complete the picture of the transport of matter from the asteroid belt. Indeed, Yarkoswsky-induced mobility of the smallest fragments generated by a catastrophic event leads them to cross a resonance after having drifted through the main belt for a period of time long enough to justify their ageing – a long journey for a future meteorite, which might well end its travels in the showcase of a museum.

Long live Yarkowsky!

Ivan Osipovich Yarkowsky (1844–1902) was not a scientist but a civil engineer who occasionally conducted research following his own interest. He was the first to identify the subtle dynamical effect that now bears his name, and published his results around 1900 in a pamphlet which is now referred to as the 'Yarkowsky lost paper'. We know about its existence because the Estonian astronomer Ernst Oepik (1893–1985) duly referenced Yarkowsky's contribution in his studies on the dynamics of interplanetary matter. The subsequent developments have much to do with the interdisciplinary nature of celestial mechanics, and in particular with the work of the Italian planetary scientist Paolo Farinella (1952–2000). In a more generalised form the Yarkowsky effect could account for some unexplained perturbations of the LAGEOS geodetic satellite (see Chapter 1), and this also led to its successful application to the dynamics of small asteroids. The precise position of asteroid (6489) Golevka, obtained by the Arecibo radio telescope in Puerto Rico in 2003, has allowed us to measure the predicted 15-km Yarkowsky drift.

JURASSIC ASTEROIDS

A meteorite falling on our planet does not represent an immediate danger for life, as it is an occasional event with a very low probability of doing harm to a human being. A comet leaves a heritage in the form of beautiful showers of meteors as the dust particles ejected from the comet's nucleus vapourise in the atmosphere, sometimes leaving long bright trails. Their yearly occurrence (the most famous – the Perseids – reaches maximum activity on 12 August) is explained by the passage of our planet close to the orbit of a short-period comet, where the density of the particles is much higher.

A more substantial hazard for our planet is represented by the near-Earth objects (NEOs) – relatively small celestial bodies with a maximum diameter of some tens of kilometres (Figure 5.11 and 5.12). As previously discussed, most of them are fragments of asteroid coming from the main belt, or dead cometary nuclei with orbits dangerously close to the Earth's orbit. The change from an astronomical scale to a human scale allows a 'small asteroid' to cause a 'big catastrophe'. A 1-km wide celestial body crashing on Earth at hypersonic velocity (thousands of kilometres per hour) can cause regional to global catastrophes.

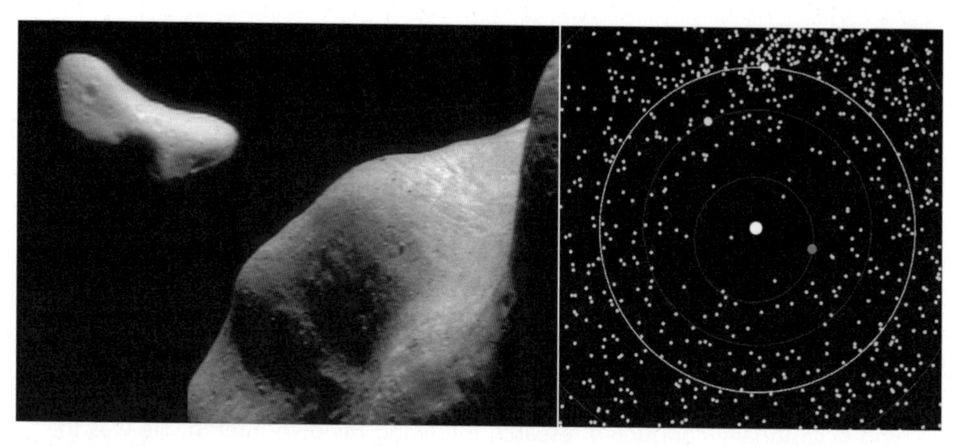

FIGURE 5.11. Two close-range views of the near-Earth asteroid Eros as imaged by the NEAR spacecraft (courtesy NASA/JHU-APL), and the positions of the known NEAs, showing that they are easily found among the orbits of the inner planets.

FIGURE 5.12. The expanding Solar System. In ancient times only the first six planets from the Sun were known, and by the end of the nineteenth century Uranus, Neptune and a few hundred asteroids had been discovered. The Solar System as we know it today is much more crowded. Three different populations of asteroids (main belt, Trojans and NEAs), the transneptunian objects and the Oort Cloud of comets have joined the family portrait, while celestial mechanics has unveiled the dynamical routes for their mobility within the Solar System.

The danger is real, as demonstrated by the events which took place in the region of Tunguska (Siberia) in June 1908, when a terrifying explosion was heard. The first scientific expedition, reaching the zone of the disaster many years later, found a desolate scene: more than 40,000 uprooted and flattened trees, and an

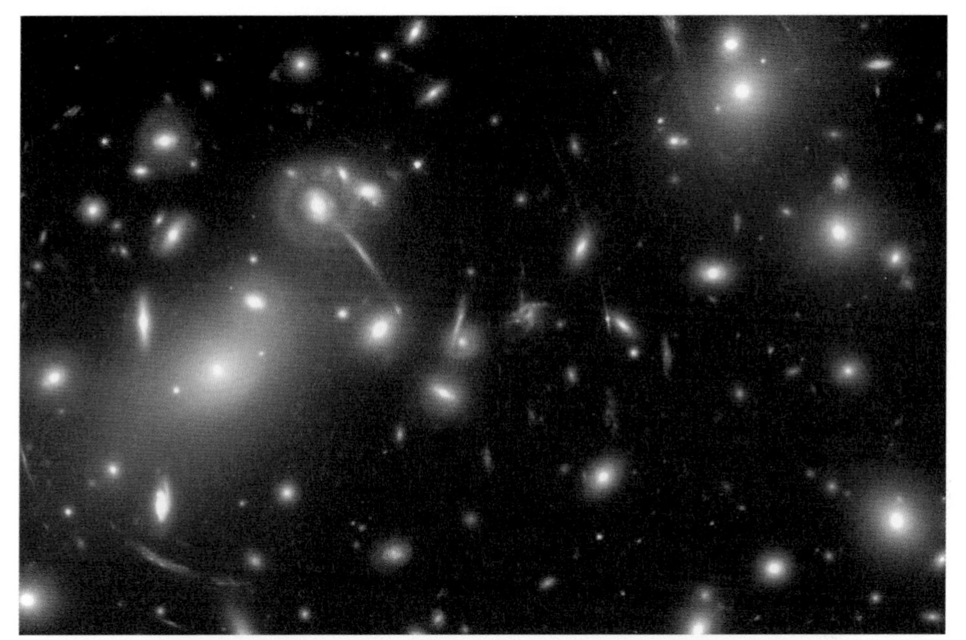

Plate 1. The galaxy cluster Abell 2218. It is located in the constellation of Draco, and lies at a distance of about 2 billion light-years. This Hubble Space Telescope image shows the gravitational lensing effect caused by the huge mass of the cluster, whereby the light of the stars behind is deflected and distorted. (Courtesy NASA, ESA, Richard Ellis (Caltech) and Jean-Paul Kneib (Observatoire Midi-Pyrenees, France), with thanks to NASA, A. Fruchter and the ERO Team (STScl and ST-ECF).

Plate 2. The beautiful Whirlpool galaxy (M51) shows its spectacular spiral arms. (Courtesy NASA, ESA, S. Beckwith (STScl), and the Hubble Heritage Team (STScl/AURA), with thanks to N. Scoville (Caltech) and T. Rector (NOAO).)

Plate 3. A wonderful image of the Orion nebula – a region of star formation – obtained by the Hubble Space Telescope. (Courtesy NASA, ESA, M. Robberto (Space Telescope Science Institute, ESA) and the Hubble Space Telescope Orion Treasury Project Team.)

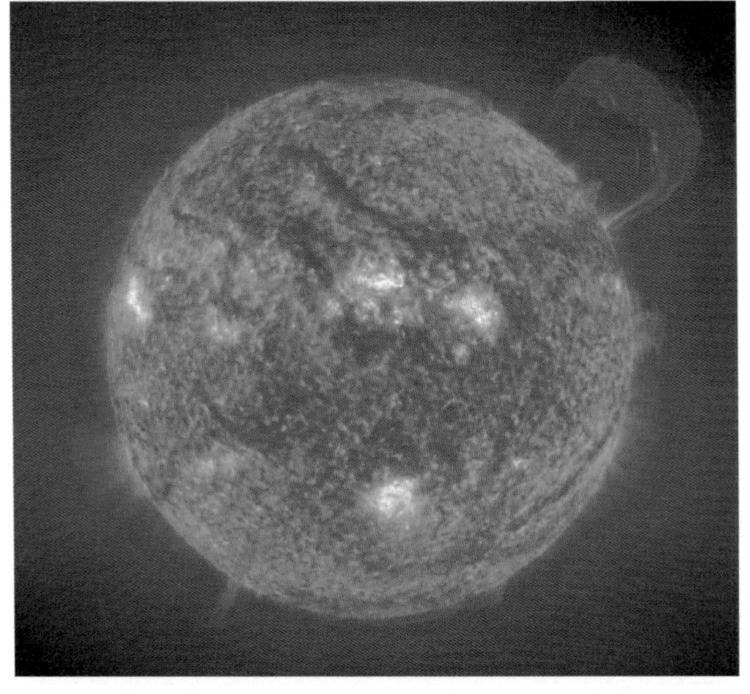

Plate 4. Solar eruptions and prominences imaged by the Extreme Ultraviolet Imaging Telescope on 14 September 1999. (Courtesy SOHO mission, NASA, ESA.)

Plate 5. A mosaic of images of Mercury obtained by Mariner 10 from a distance 125,000 miles. The tiny, brightly rayed crater (just below centre top) was the first recognisable feature on the planet's surface, and was named in memory of astronomer Gerard Kuiper (a Mariner 10 team member). (Courtesy NASA, Jet Propulsion Laboratory.)

Plate 6. An ultraviolet image of Venus's clouds, obtained by Pioneer Venus in 1979. (Courtesy NASA).

Plate 7. The Earth, photographed by the crew of Apollo 17 during their journey to the Moon. (Courtesy NASA.)

Plate 8. The Apollo 11 Lunar Module with Earthrise in the background; 11 July 1969. (Courtesy NASA, NASA History Office and the NASA JSC Media Services Center.)

Plate 9. The largest crater in this picture of the lunar surface is Daedalus. Located near the centre of the far side of the Moon, its diameter is about 58 miles (93 km). This image was taken from lunar orbit by the crew of Apollo 11. (Courtesy Apollo 11, NASA.)

Plate 10. A spectacular image of Mars (Courtesy NASA, ESA, and the Hubble Heritage Team (STScl/ AURA).)

Plate 11. A panorama of Mars obtained by the Mars Exploration Rover Spirit during its excursion toward the Columbia Hills on 12–13 March 2004. (Courtesy NASA, Jet Propulsion Laboratory, Cornell.)

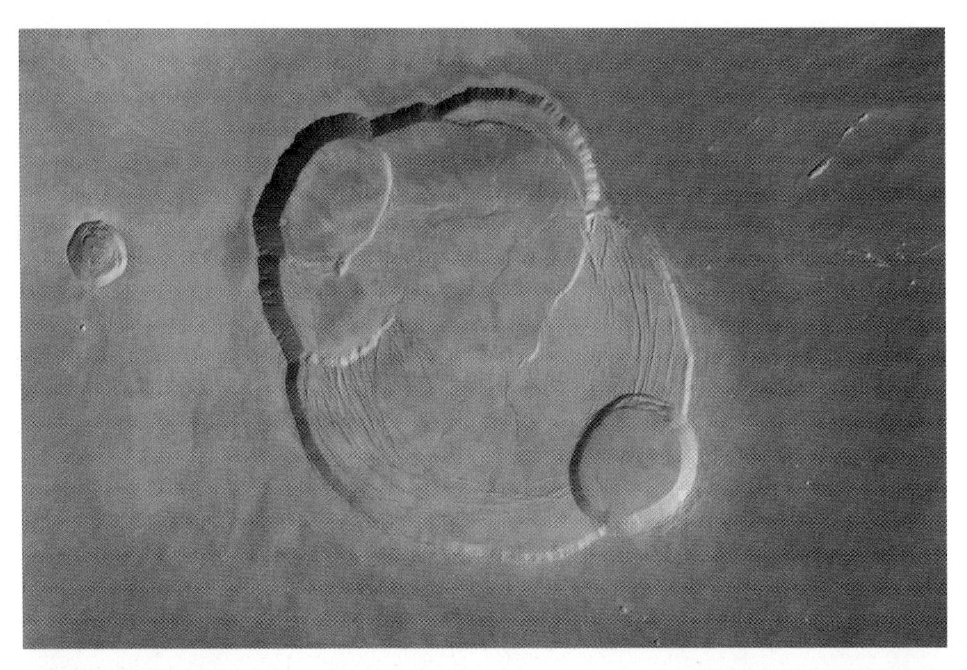

Plate 12. The highest volcano in the Solar System is Mons Olympus on Mars, with an average height of 22 km. This image was obtained by Mars Express on 21 January 2004. (Courtesy ESA, DLR, FU Berlin (G. Neukum).)

Plate 13. As it swooped past the south pole of Saturn's moon Enceladus on 14 July 2005, the Cassini spacecraft acquired high-resolution views of this puzzling ice world. From afar, Enceladus exhibits a bizarre mixture of softened craters and complex, fractured terrains. In this large mosaic, twenty-one narrow-angle camera images have been arranged to provide a full-disk view of the anti-Saturn hemisphere of Enceladus. It is a false-color view that includes images taken at wavelengths from the ultraviolet to the infrared. (Courtesy NASA, Jet Propulsion Laboratory, Space Science Institute.)

Plate 14. The Galileo spacecraft provided us with this superb image of the asteroid Ida and its tiny satellite Dactyl. (Courtesy NASA, Jet Propulsion Laboratory.)

Plate 15. A MERIS reduced resolution mode image of Hurricane Rita in the Gulf of Mexico. (Courtesy ESA.)

Plate 16. Jupiter's Great Red Spot is an anticyclonic storm, comparable to the most severe hurricanes on Earth. This image was obtained by Voyager 1. (Courtesy Voyager 1, NASA, Jet Propulsion Laboratory.)

Plate 17. Jupiter, its moon Io, and Io's shadow. (Courtesy NASA, Jet Propulsion Laboratory, University of Arizona.)

Plate 18. A beautiful image of Saturn obtained by the Cassini–Huygens spacecraft four months before its arrival at the planet. (Courtesy NASA, Jet Propulsion Laboratory, Space Science Institute.)

Plate 19. Hubble Space Telescope images of the apparent inclination of Saturn's rings throughout 1996–2000. (Courtesy NASA and The Hubble Heritage Team (STScI/AURA), with thanks to R.G. French (Wellesley College), J. Cuzzi (NASA/Ames), L. Dones (SwRI), amd J. Lissauer (NASA/Ames).)

Plate 20. Saturn's ring system, imaged by Voyager 2. The colours have been enhanced. (Courtesy NASA, Jet Propulsion Laboratory.)

Plate 21. Saturn and its moons Dione (foreground), Tethys and Mimas (right), Enceladus and Rhea (left), and Titan (top right). A montage of pictures obtained by Voyager 1 in 1980. (Courtesy NASA, Jet Propulsion Laboratory.)

Plate 22. The Galileo spacecraft acquired its highest-resolution images of Jupiter's moon Io on 3 July 1999, during its closest pass to Io after orbit insertion in late 1995. Most of Io's surface has pastel colors, punctuated by black, brown, green, orange and red units near the active volcanic centres. (Courtesy Galileo Orbiter, NASA, Jet Propulsion Laboratory, Planetary Image Research Laboratory, University of Arizona.)

Plate 23. Few sights are more impressive than a bright comet hanging in the sky just after sunset or shortly before dawn. This view of comet Hale-Bopp was taken on 9 March 1997 at 0500 GMT. The comet's two principal tails are clearly shown. (Courtesy Glyn Marsh and *The Astronomer.*)

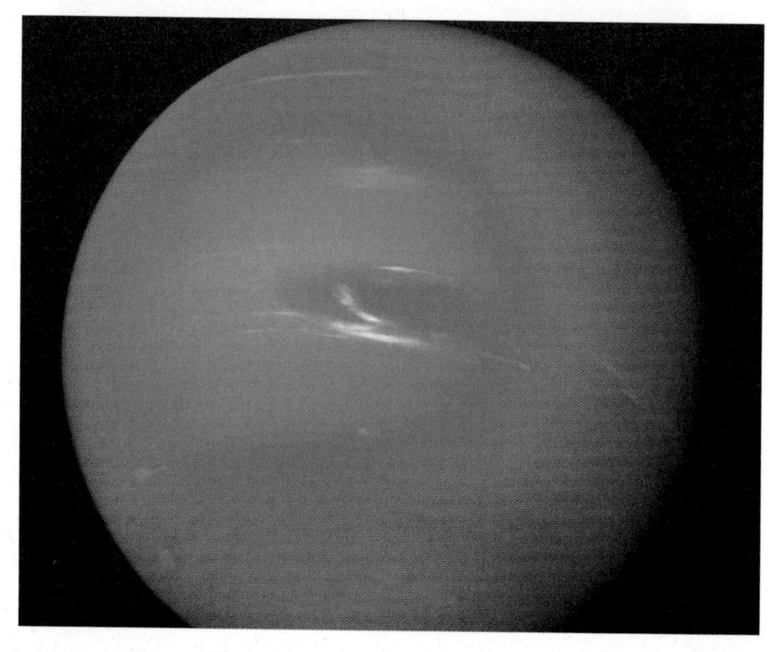

Plate 24. An image of Neptune by Voyager 2, taken at a range of 4.4 million miles from the planet, 4 days 20 hours before closest approach. The picture shows the Great Dark Spot and its companion bright smudge. On the west limb, the fast-moving bright feature called Scooter and the little dark spot are visible. (Courtesy Voyager 2, NASA, Jet Propulsion Laboratory.)

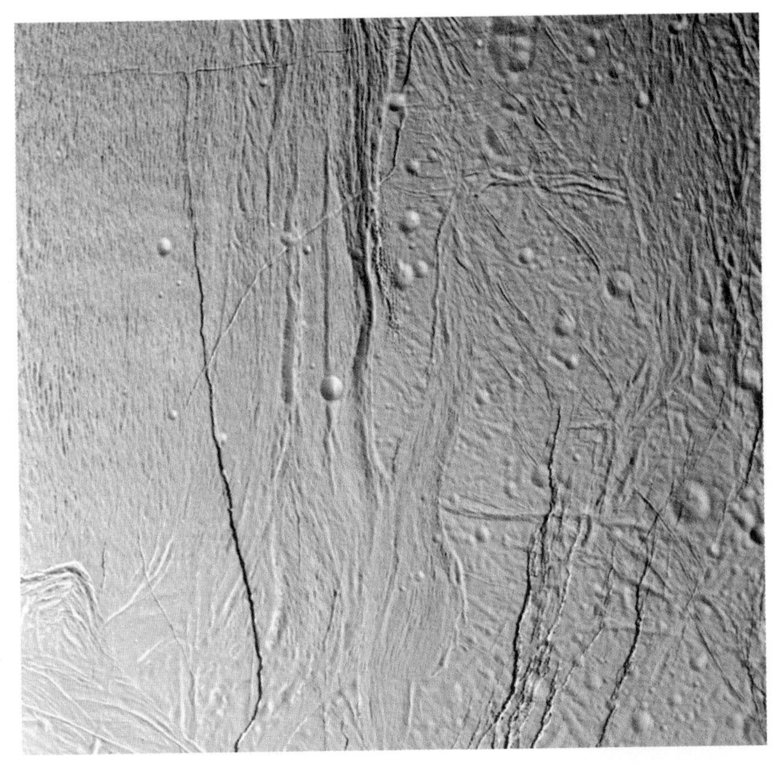

Plate 25. The icy surface of Saturn's satellite Enceladus, imaged by the Cassini spacecraft. (Courtesy NASA, Jet Propulsion Laboratory, Space Science Institute.)

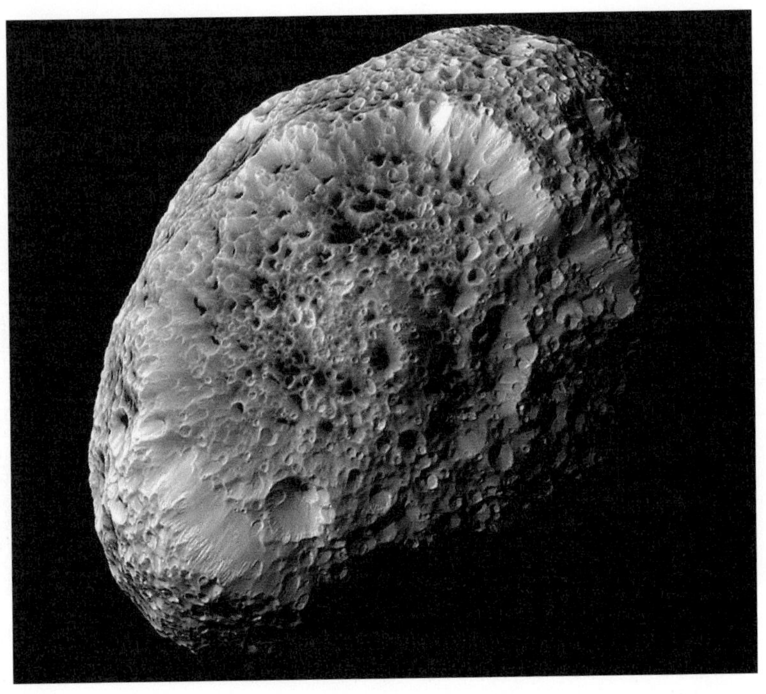

Plate 26. Hyperion – a small and irregular satellite of Saturn – imaged by the Cassini spacecraft from a distance of about 62,000 km. (Courtesy NASA, Jet Propulsion Laboratory, Space Science Institute.)

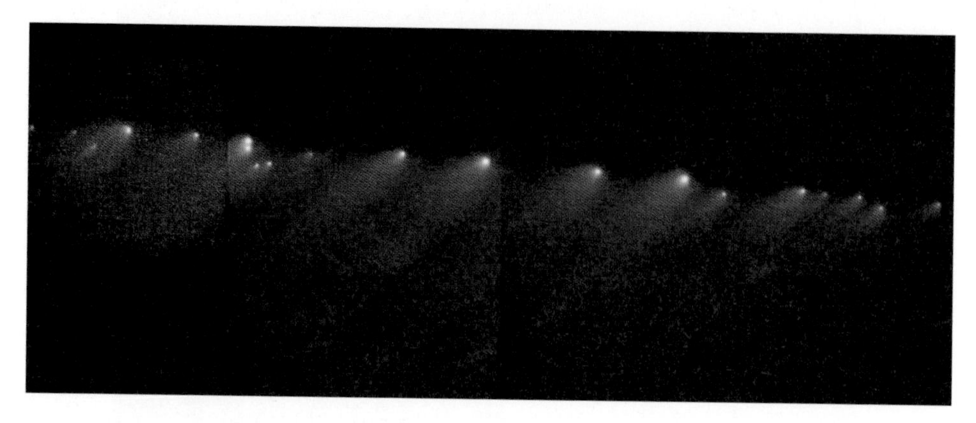

Plate 27. The fragments of comet Shoemaker–Levy 9, which collided with Jupiter in July 1994. (Courtesy Dr Hal Weaver and T. Ed Smith (STScl), and NASA.)

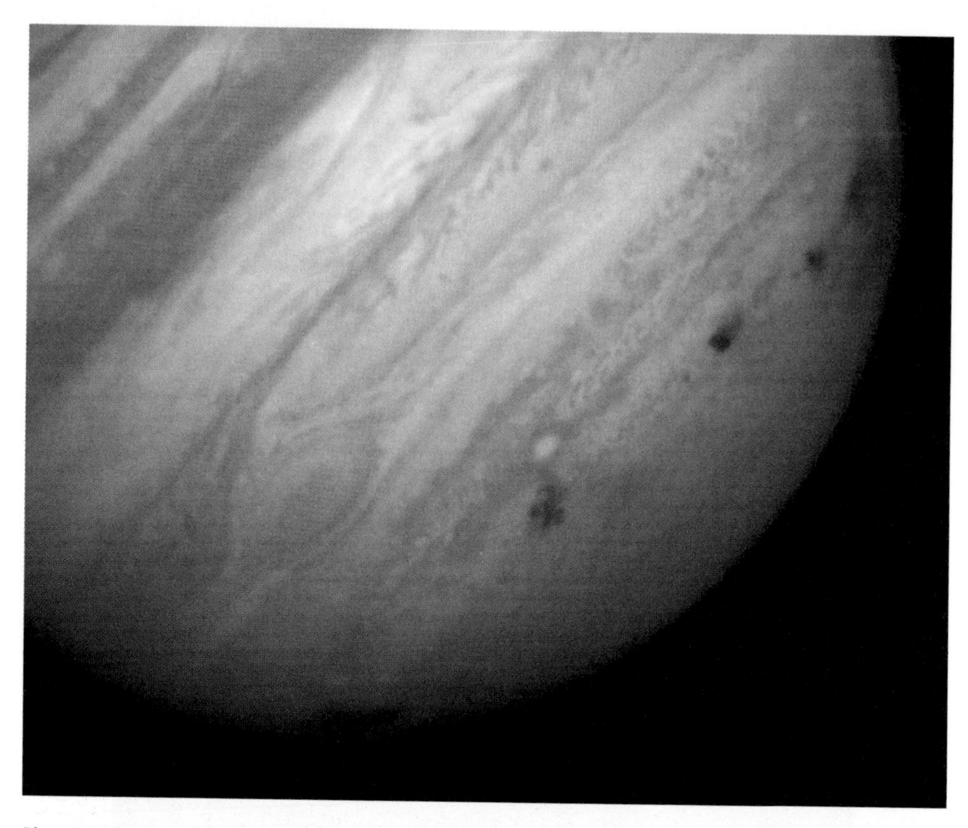

Plate 28. The scars left by comet Shoemaker–Levy 9 in the atmosphere of Jupiter, imaged in July 1994. (Courtesy NASA and the Hubble Space Telescope Comet Team.)

Plate 29. Part of the Sagittarius star-cloud in the heart of the Milky Way. (Courtesy NASA and the Hubble Heritage Team (STScI/AURA).)

Plate 30. The Sombrero galaxy (M104) at the southern edge of the Virgo cluster of galaxies. (Courtesy NASA and the Hubble Heritage Team (STScI/AURA).)

Plate 31. The spectacular end of a star: the supernova remnant LMC N49, located within the Large Magellanic Cloud. (Courtesy NASA and the Hubble Heritage Team (STScl/AURA), with thanks to Y.-H. Chu (UIUC), S. Kulkarni (Caltech) and R. Rothschild (UCSD).)

Plate 32. The strong interplay of the galaxies NGC 2207 and IC 2163 will possibly lead to their merging within a few billion years. (Courtesy NASA and the Hubble Heritage Team (STScl/AURA).)

increase of the genotypic mutability of pine trees in the region of the catastrophe – but no clear signs of an impact crater or of meteorite fragments. The lack of evidence for a meteoritic impact gave rise to exotic explanations, such as matter–antimatter annihilation, an encounter with a small black hole, and, inevitably, a flying saucer out of control. It is now thought that a small stony asteroid between 30 and 60 metres in diameter exploded before reaching the ground, at a height of about 5–10 km, vapourising almost completely and releasing energy of approximately 10 megatons – the equivalent of an intermediate-power nuclear bomb. From a statistical point of view, a Tunguska-like event happens, on average, every few centuries. Global catastrophes are triggered only by collisions with bigger objects, with much lower probability, thus being separated by million years or more.

Yet statistics, by its own nature, is not reassuring, and the study of NEAs is attracting an increasingly large audience. The astronomical community has set up dedicated programmes for NEA discovery and follow-up, with the goal of cataloguing all potentially hazardous objects. Civil defence has included asteroid impact in the list of possible natural disasters, while the public interest is high on this matter because of the 'near-miss' announcements often appearing in the press. The good news is that space agencies are beginning to develop realistic plans for protecting the Earth by destroying or deflecting an asteroid on a collision course with our planet.

Dinosaurs did not have this chance. Their extinction, some 65 million years ago, is thought to be a direct consequence of a cosmic impact. The connection between dinosaurs and asteroids began as a working hypothesis in 1978, when in the region surrounding the city of Gubbio (Italy) geologists found an anomaly in the composition of the terrain dating back to the Cretaceous–Tertiary (K–T) boundary. An anomalous concentration of iridium, which is present in larger quantities on celestial bodies than on the surface of the Earth, was interpreted as a sign that a major impact occurred on our planet at roughly the same epoch as when the dinosaurs' mass extinction event took place.

Initially this appeared to be a remarkable coincidence, and nothing more. Dinosaurs ruled the Earth for more than 100 million years and were spread around the world. Bones and skeletons have been found on almost every continent from Europe to Australia and from Asia to America. But how could they suffer so much from a single impact, however large? The answer is that beyond a certain size the consequences of an impact cease to be regional and act on a global scale. The release in the atmosphere of the large amounts of dust generated by an impact dramatically altered the daily heat influx from the Sun, thus leading to a sudden climate change everywhere on the planet. Evolution by natural selection does not work at such short notice, and the result in the case of the K–T boundary was that as many as 70% of the life-forms, including the dinosaurs, were wiped out.

As in any good thriller in which the victim is known, the bullet has been found but the killer is not known, the final proof of the existence of a dinosaur-killer asteroid has been missing for some time. The big crater – at least 100 km across – that such a large impact should have left behind was nowhere to be seen, and there was little hope of ever discovering it. The Earth is geologically very much alive. Volcanism, earthquakes, weathering and erosion are continuously changing its surface, and none of the few impact craters still recognisable had the required size and age. However, the quest for the missing crater eventually succeeded due to modern remote sensing techniques. Buried beneath the sea floor of the north coast of Mexico, facing the Yucatan Peninsula, the edges of a 180-km crater were identified. As a tribute to the ancient Maya civilisation which once ruled the land, the crater was named Chicxulub – the 'devil's tail'.

This scenario is intriguing both from an astronomical point of view and concerning the origin of our own species. The disappearance of the dinosaurs is considered a key event which allowed mammals such as lemurs to evolve toward hominids. It seems that we are now fearing the very same astronomical event which paved the way for intelligent life to appear on Earth.

The frightening interplay between asteroids and dinosaurs represents only one of the possible end states for a wandering celestial body. In general, the dynamical evolution of the asteroid population can be considered a slow chaotic diffusion process from the main belt towards the inner Solar System. Close encounters and collisions with the terrestrial planets do indeed occur, but it is not the worst that can happen. The Sun – with its 700,000 km radius and a mass 1,000 times larger than that of all the planets combined, represents the ultimate destination. Extensive numerical experiments have shown that the most probable fate for a chaotic asteroid is that it will fall into this immense solar furnace.

6

Singularities, collisions and threatening bodies

The eternal silence of these infinite spaces frightens me.

Blaise Pascal

Celestial bodies collide much more frequently than was thought in the past. The impact craters on the surface of planets and satellites, the long bright trails left by fireballs in the Earth's atmosphere, and on a broader scale the slow merging of galaxies, reveal phenomena that continuously shape the structure of the Universe. Yet a collision, from a strictly mathematical point of view, has a deeper meaning involving peculiar entities such as zero and infinity. The whole matter has recently become of great interest because of the threat posed by the NEO population – celestial bodies potentially on collision trajectories with the Earth. The good news is that for the first time in the history of mankind, space-based mitigation strategies for avoiding a catastrophic impact on our planet can be implemented. The story begins with a complicated interplay of fascinating terms: singularity, collisions and infinity.

FROM ZERO TO INFINITY

Celestial mechanics is ruled by gravitation. All bodies attract each other, and the attractions are stronger with larger mass and fade as the distance between the bodies increases. The mathematical formulation of this fundamental principle, explaining in detail how gravitation works, is given by Newton's law (introduced in Chapter 1), which states that the intensity of the force F acting between any pair of massive bodies M_1 and M_2 is directly proportional to the product of their masses and inversely proportional to the square of their distance d:

$$F = -G \frac{M_1 M_2}{d^2}$$

In general, the finite size of celestial bodies prevents their relative distance from being exactly zero. A collision between two spherical bodies occurs if their centres are separated by a distance less than the sum of their radii. If this were not so, the introduction of the value $d = 0$ in the denominator of the expression of the force F implies that its intensity reaches infinity. As a consequence, Newton's equation ceases to be valid and the description of the motion fails. If this happens, mathematically speaking we are dealing with a *singularity* – an 'event' (in a broad sense) which requires specific treatment.

Infinity

A simple argument can be followed for obtaining evidence that the result of the division of a number by zero is infinity. Let us take the number 1 (although the same applies to any other number) and start dividing it by increasingly smaller quantities: 1/0.1 = 10, because, converting decimal numbers into fractions, 0.1 = 1/10, and it is therefore necessary to add 0.1 ten times to obtain unity again. Similarly, when 1 is divided by 0.01 the result is 100, while 1/0.001 = 1,000. It can be clearly recognised that adding extra decimal zeros to the divisor, which corresponds to using smaller and smaller numbers, the result grows larger and larger. Bringing this example to the limit, it can be said that when the divisor equals zero, then an infinitely large number is obtained. This is why division by zero is denoted with the symbol for infinity: ∞.

To play safe it should be admitted that is not always pleasant to work with mathematical singularities, as they usually form an obstacle which forces theories to become more complicated. However, a proper treatment of singularities is essential, because they appear in many fields of science, from mathematics to biology and from the atomic structure of matter to cosmology.

In the mathematical context, singularities appear under different guises, in addition to the algebraic division by zero. In the delicate field of complex variables they bear high-sounding names: poles, branch points and essential singularities. Nodes, cusps and isolated points pertain to the study of algebraic curves.

The idea of a singularity as the basis of a cosmological model was already present in Stoic philosophy, and was denoted with the term *ekpyrosis* (conflagration). The modern Big Bang theory of the origin of the Universe sees a similar explosion in which stars, planets and galaxies emerged from a singularity, when space, energy, time and matter were concentrated in a point with infinite density.

The Solar System has experienced (and still experiences) events which can be considered as singularities, since they are characterised by a sharp transition from a given dynamical and/or physical state to a completely different state. The cratered surfaces of rocky satellites and planets are all of what is left of the many small bodies that completely disappeared as individual objects due to hypervelocity impacts. Asteroid families (introduced in Chapter 5) witness the partial or complete disruption of the colliding parent bodies (see Figure 6.1).

In celestial mechanics it is difficult to study the orbital evolution of a system even in the neighbourhood of a singularity, as extensively discussed in the

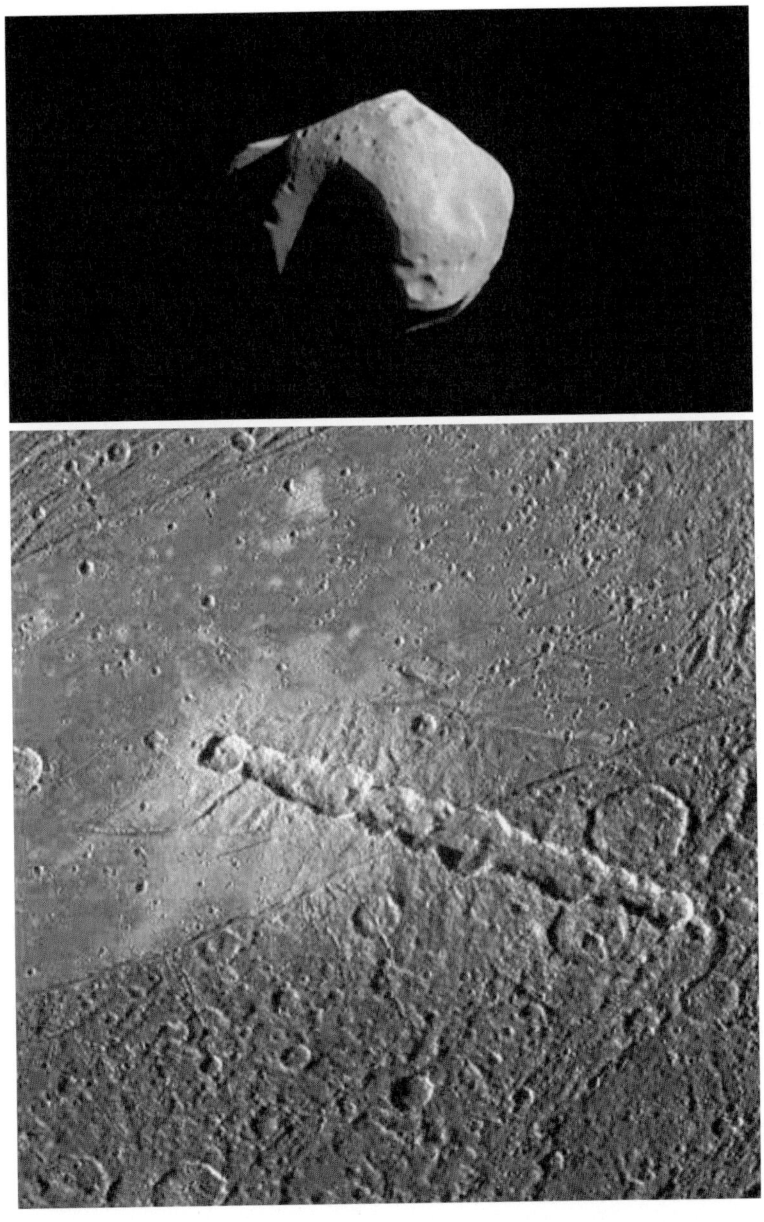

FIGURE 6.1. Impact craters are everywhere in the Solar System, on small celestial bodies such as the main-belt asteroid Mathilde (top) (Courtesy NASA/JHU-APL), as well as on planetary-sized icy satellites such as Ganymede, where a chain of impact craters is clearly recognisable. (Courtesy NASA/JPL.)

previous chapters, when dealing with the dynamics of a close encounter. Mathematical theories have been developed to this end, with a flourishing of ideas at the turn of the nineteenth century. In particular, T. Levi-Civita, G.D. Birkhoff, P. Kustaanheimo and E.L. Stiefel laid the foundations of the *regularisation theory*, which provides a method for studying the dynamics of two closely interacting bodies. This theory is founded upon two main components: the introduction of a suitable change of coordinates, which acts as a magnifying glass in the region of the motion where a collision takes place; and a stretching of the timescale, with the introduction of a 'fictitious time', resulting in a sort of 'slow motion' at collision.

The implementation of regularisation theory on powerful computers allows a detailed investigation of the dynamics in the close proximity of a singularity.

COLLISIONS IN THE SOLAR SYSTEM

Collisions are not simply destructive. Far back in the history of the Solar System the frequent collisions of protoplanets with the crowd of planetesimals orbiting the interplanetary space contributed significantly to their growth until they reached their present size. A giant collision is also the most credited hypothesis on the origin of the Moon (as described in Chapter 7). Comets have delivered water (and perhaps, some have theorised, primitive life forms) throughout the Solar System by impacting on other celestial bodies. It has been proposed that most of the water in the oceans was produced by comets impacting the Earth early in its history. A similar origin has been proposed for the large reservoirs of ice which are expected to be found inside permanently shadowed lunar craters. They could greatly contribute to a stable human presence on our satellite by presenting precious *in situ* resources.

The populations of celestial objects for which collisions act as a major evolutionary process can be outlined as follows:

Main-belt asteroids The size distribution of the rocky bodies orbiting in the region between Mars and Jupiter is clearly the result of intense collisional evolution. Only the largest members of the population, such as Ceres and Vesta, have an almost spherical shape, while the vast majority are irregularly shaped bodies ranging from a few hundred kilometres to metre-sized objects. About 100,000 main-belt asteroids have so far been catalogued, but the entire population is estimated to be in the order of about a million objects with a diameter of more than 1 km.

Near-Earth asteroids (NEA) Small asteroids on orbits that closely approach that of the Earth and are therefore presently (or are likely to become in the future) hazardous Earth-crossing bodies. Their origin can be traced back to the asteroid main belt as small fragments delivered to the inner Solar System on chaotic orbits (see Chapter 5). Their chaotic dynamics leads them to eventually collide with a planet or fall into the Sun. A few thousand NEAs are presently known, while the whole population is estimated to be in the order of 30,000 – 300,000 objects larger than 100 metres across.

Short-period comets Kilometre-sized icy bodies with revolution periods less than 200 years and characterised by a wide range of orbital regimes such as high eccentricity and inclination as well as almost circular orbits close to the ecliptic. Their chaotic dynamics is controlled by strong perturbations and close encounters with the outer giant planets, thus allowing collisions. Due to the relatively small number of short-period comets – only a few hundred – the hazard posed to the Earth is negligible on a short timescale.

Long-period comets Icy bodies on high eccentricity and inclined orbits. Those entering for the first time into the inner regions of the Solar System, directly from the Oort Cloud, are considered as 'new' comets. Their appearance is rare and unpredictable, and as they approach from every direction in the sky the probability of impacting an inner planet is low. A significant fraction of long-period comets – the Kreutz group – have perihelia smaller than the radius of the Sun, and therefore eventually enter the Sun and are destroyed.

Edgeworth–Kuiper Belt objects (EKBO) These have undergone significant collisional evolution, and are indicated as the source region of short-period comets.

Other types of objects are also involved in collisions with the Earth:

Meteors The tiny dust particles ejected from the nucleus of a comet slowly spreads along its orbit due to perturbations (for example, solar radiation pressure), and many short-period comets are associated with complex systems of debris which cross the Earth's orbital path. When this happens a meteor shower occurs, characterised by the complete burning of the dust grains in the Earth's atmosphere, leaving long bright trails, commonly known as 'shooting stars'.

Meteorites If a celestial body is large enough to survive high-velocity entry into the atmosphere of our planet, but small enough to not be dangerous, it can reach the ground. These objects can be considered as the lower end (diameter less than 50 metres) of the near-Earth asteroid population, as confirmed by the similarities between their composition and that of the main-belt asteroids.

Near-Earth objects (NEO) The NEO population encompasses all celestial bodies that due to their size and dynamics may be significantly dangerous if a collision with our planet occurs in the near future. The NEO population is composed of all NEAs and comets with perihelia less than 1.3 AU, and by a few man-made pieces of 'space junk' that have escaped into orbits close to that of the Earth.

The above populations are connected by evolutionary patterns, and intruders are often found. Apart from the aforementioned spaceways linking main-belt asteroids with NEAs, and EKBOs with short-period comets, it is sometimes difficult to distinguish between a comet and an asteroid, both from an observational and a dynamical point of view. It is widely believed that many short-period comets are hidden inside the NEA population as dead cometary nuclei, exhibiting no activity because of the thick crust formed after repeated passages close to the Sun. Comets can also travel the other way around the dynamical paths connecting the inner planetary region with the asteroid main belt, and begin orbiting between Mars and Jupiter as typical members of the asteroid population (no cometary activity is displayed because of the large distance from the Sun).

Vesta and the Vestoids

The large asteroid (4) Vesta (500 km in diameter) represents a unique case in many respects. The peculiar 'colour' of the light coming from its surface indicates a basaltic composition, which is well known to be an outcome of volcanic activity. Moreover, a peculiar class of meteorites displays similar composition, thus indicating their origin as fragments ejected in space after a major impact on the asteroid's surface. An additional indication was given by Rick Binzel and Shui Xu, of the Massachussets Institute of Technology, who were able to identify several small asteroids that could be considered as the largest fragments blasted away during the same impact event, independently orbiting inside the main belt. Images of Vesta obtained by the Hubble Space Telescope in 1993 show a roughly spherical celestial body (Figure 6.2) with a large impact crater (with a diameter almost as large as Vesta itself) located near the south pole, providing the required observational evidence of a past collision.

The existence of this collisional family is now commonly accepted, and the members of the family are termed Vestoids.

ASTROBLAMES

Today cosmic impacts are occasional events considered from two aspects: the enjoyment of spectacles such as the periodic occurrence of meteor showers and the 1994 collision of comet Shoemaker–Levy 9 with Jupiter; and the fear of the non-zero probability that at some time in the future an asteroidal impact could cause regional disasters or even wipe out life on our planet.

Almost 100 tons of interplanetary matter – mostly dust grains – enter the Earth's atmosphere every day, and it has been estimated that each year approximately 500 meteorites weighing about 100 grammes fall on every million square kilometres of the Earth's surface. Luckily the collision with celestial bodies sufficiently large to leave a tangible sign of their arrival on the Earth's surface is much less frequent. The craters generated by these impacts have a roughly circular shape, whatever the slope of the incoming trajectory, because they are produced by veritable explosions in which energy is not provided by chemical or nuclear reactions but by celestial mechanics. The extremely high relative velocity at which the body hits the ground results from the different

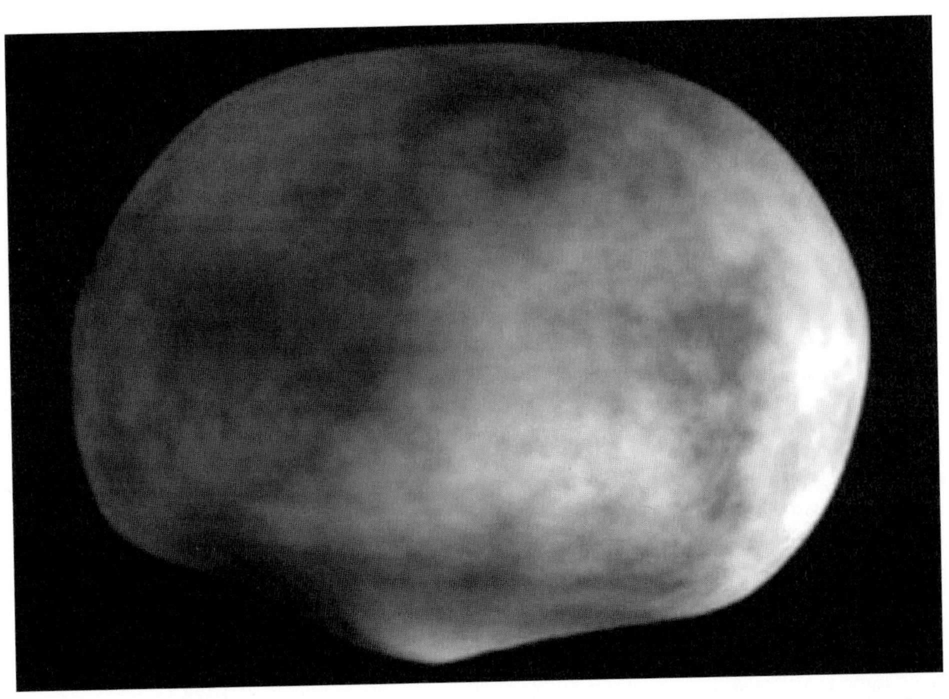

FIGURE 6.2. A Hubble Space Telescope image of asteroid Vesta. A large impact basin and a central peak characterize the South Polar region.

orbital motion of the Earth and of the impactor, enhanced by the 'gravitational focusing' of our planet – the acceleration caused by the non-negligible mass of the Earth.

A list of the major impacts suffered by the Earth is given in Table 6.1, which provides an estimate of the time at which the impact presumably occurred and of the diameter of the remnant crater or depression.

Table 6.1. Major impacts on the Earth.

	Diameter (km)	Age (years)
Vredefort, South Africa	300	2 billion
Sudbury, Ontario (Canada)	250	1.85 billion
Chicxulub, Yucatan (Mexico)	170	65 million
Popigau, Siberia (Russia)	100	35.7 million
Manicouagan, Quebec (Canada)	100	214 million

A list of relevant impact craters surviving the continuous resurfacing activity of our planet is shown in Table 6.2. Possibly the most famous of them – Barringer Crater (also known as Meteor Crater) – has a diameter of about 1.2 km and reaching a depth of 170 m. Discovered in 1891, it is a large almost circular depression in the northern Arizona desert. Its cosmic origin was recognised only after a long debate within the scientific community, because of the possible alternative explanation as a volcanic event. Eventually, the convincing evidence was produced by Daniel Moreau Barringer, who pointed out the presence of a large quantity of dust having a typical meteoritic composition (nickel–iron) scattered around the crater, and the absence of nearby volcanic formations. It is now believed that the crater was created 50,000 years ago by an asteroid with a diameter of about 30 metres.

Table 6.2. The most relevant impact craters on the Earth.

	Diameter (metres)	Year of discovery
Meteor Crater, Arizona	1,265	1871
Wolf Creek, Australia	850	1947
Henbury, Australia	200 × 110	1931
Boxhole, Australia	175	1937
Odessa, Texas	170	1921
Wabar, Arabia	100	1932
Oesel, Estonia	100	1927
Campo del Cielo, Argentina	75	1933
Dalgaranga, Australia	70	1928
Sichote-Alin, Siberia	28	1947

The Wabar crater, located in the Rub'al-Khali desert in southern Saudi Arabia, is composed of three main circular depressions – the largest measuring about 100 metres. It was discovered in 1932 by the British explorer Harry St John 'Abdullah' Philby, while he was looking for the legendary town of Ubar. Again, the evidence was provided by nickel–iron rocks spread around the crater rim. The impactor weighed about 3,500 tons, and hit the Earth at a low angle with an estimated speed of 40,000–60,000 km/h. During the atmospheric descent it was probably disrupted into several pieces, producing a number of craters.

Located on the opposite side of the planet, the Rio Cuarto craters in Argentina are attributed to one or more celestial bodies impacting the Earth on an almost grazing trajectory. They date from about 10,000 years ago.

A peculiar situation is found in Bavaria, Germany. The large depression named Noerdlingen-Ries is composed of two craters. The larger of them has a diameter of 25 km and is about 240 m deep, while the smaller is 2.5 km wide and about 100 m deep. They were believed to be of volcanic origin until in 1960 Eugene Shoemaker and Edward Chao revealed their true nature as twin cosmic impacts dating from 14.8 million years ago. The city of Noerdlingen is built inside the main crater, and its church is built completely of rock created by the impact.

It is not always easy to trace back the 'astroblames' – the celestial bodies responsible for the 'crime' – especially when their dimensions are small. Icy bodies tend to vapourise in the atmosphere, while rocky objects are often fragmented into several pieces. When a meteorite fall is observed – recognisable due to its unusual brightness and by its slower speed compared with a meteor – astronomers dress up as explorers and scan the countryside in search of precious extraterrestrial samples.

The largest meteorite found on the Earth – Hoba-West, in south-west Africa – remains in the location where it landed in prehistoric times. It weighs about 61 tons (see Table 6.3).

Table 6.3. Some of the largest meteorites recovered.

	Place of discovery	Weight (tons)
Hoba-West	Africa	61
Ahnighito	Greenland	30.9
Bacuberito	Mexico	27.4
Mbosi	Tanganyika	26.4
Agpalik	Greenland	20.4
Armanty	Mongolia	20
Willamette	USA	14
Chupaderos	Mexico	14
Campo del Cielo	Argentina	13
Mundrabilla	Australia	12
Morito	Mexico	11

NEAR-EARTH ASTEROIDS

When, a little more than a century ago, the orbit of the newly discovered asteroid 1898 DQ was determined, astronomers were caught by surprise. Instead of displaying the expected mean distance from the Sun between the orbit of Mars and Jupiter, it moved on an elongated trajectory with a perihelion remarkably close to 1 AU. Was it an exception, or was it the first member of a new population of small bodies which could dangerously approach our planet? Time has shown that 433 Eros (as the peculiar object was named) is one of the largest near-Earth asteroids (about 30 km in diameter) – and not even the most potentially dangerous. For example, asteroid (4179) Toutatis, discovered in 1989, approaches the orbit of the Earth roughly every four years. In September 2004 it approached the Earth to within 1.5 million km – only four times farther away than the Moon. Repeated close encounters with our planet afterwards allowed us to carry out high-resolution radar observations of the asteroid with the Arecibo radio telescope. Refined data-processing produced a radar image of Toutatis (Figure 6.3) showing a highly irregular body of 2–3 km in diameter – an object which, if it hits the Earth, would lead to consequences on a global scale.

Orvinio, 31 August 1872

The fall of a meteorite is a remarkable event, rarely witnessed by human eyes because of the relatively stringent constraints posed by location and timing. The appearance of an extremely bright fireball in the skies above Rome in the early morning of 31 August 1872 represented a unique opportunity, due to the densely populated region and the prompt interest of distinguished scientists. From the newspapers of that time we know that exhaustive investigations were carried out by the Director of the Vatican Observatory, the celebrated astronomer Angelo Secchi, supported by Professor Michele Stefano De Rossi, a renowned Italian geologist.

The meteorite came from the south, passed over the countryside east of Rome, and entered a region characterised by hills and mountains. There it repeatedly exploded, throwing fragments everywhere – some of which were later found. De Rossi saw the spectacular fireball from the small town of Rocca di Papa, near Rome, and immediately went searching for witnesses and meteoritic samples. His account of what he found in the region is breathtaking. Conic-shaped holes about 40 cm deep excavated by the smaller fragments could be easily found; a small building used for storing hay was on fire; and people were deeply frightened by the strange loud thunder which made window glass tremble, and by the strange burning stones whistling down. A shepherd fainted with terror, and the local doctor was repeatedly called for similar cases. De Rossi eventually succeeded in gathering several pieces of the object that had caused so much turmoil. They had been dispersed over a region several kilometres wide, and the largest fragment was found near Orvinio, a small town east of Rome. This fragment weighed 735 grammes, and had formed a crater about 60 cm in diameter. Here is what the 'Arciprete di Orvinio', Don Valentino Valentini, declared: 'The phenomenon was absolutely amazing, villains fell on the ground and some fainted because of the fiery rocks falling beside and which they believed were lightnings descending from a clear sky'

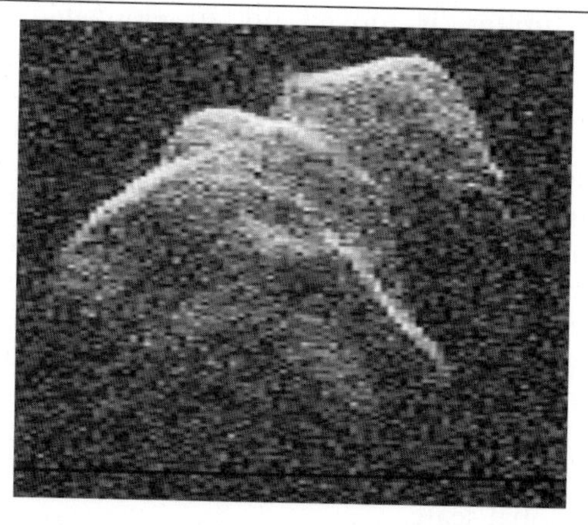

FIGURE 6.3. A radar image of asteroid Toutatis obtained by the Arecibo radio telescope.

Lost and found

Eros was discovered from the Urania Sternwarte Berlin on the night of 13 August 1898 (Figure 6.4), and the discovery was announced by Gustav Witt in *Astronomiche Nachrichten* and the *Astronomical Journal.* According to a study by Hans Scholl (Observatory of Nice) and collaborators, the discovery of Eros was not that simple. The first clue is provided by a report to the general assembly of the Société Astronomique Française on the session held on 5 October 1898, in which the circumstances of the discovery are described: 'It is certain that the planet 1898 DQ was observed in Nice on August 13, the same day as in Berlin. In order to complete his observation of August 13, Monsieur Witt made observations on August 14 and 15, which were respectively a Sunday and a holiday. I suppose that Monsieur Charlois had to postpone his observations until Tuesday, August 16, and that was sufficient for him to lose the merit of an important discovery.'

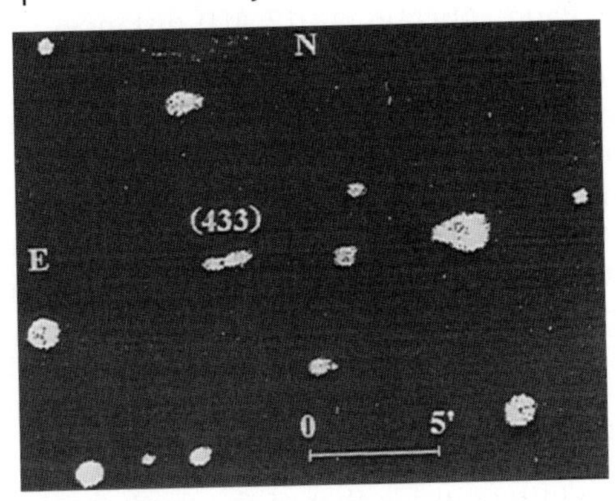

FIGURE 6.4. Gustav Witt's original discovery plate of asteroid (433) Eros.

But who was Auguste Charlois, and why did he supposedly postpone his observations? At the end of the nineteenth century Charlois was a leader in discovering asteroids, a pioneer in the emerging field of astronomical photography, and an experienced and regular observer of minor planets. How could his alleged negligence prevent him from discovering Eros? Does he deserve such a long-lasting bad reputation? In the year 2000, when the spacecraft NEAR first reached Eros, some French newspapers still reported the story of Charlois' missed discovery.

From the careful investigation mentioned above it appears that on 13 August 1898 Charlois was disadvantaged in several respects – and not through his own fault. First of all he had a mechanical problem with his telescope (as duly reported in his logbook), which resulted in the stars appearing as parallel trails, instead of dots, on his photographic plates, thus obscuring the trail produced by the motion of an asteroid. Neither was the weather favourable. Meteorological records show that it was a stormy weekend in Nice, while in Berlin the Sun was shining and the nights were clear. Lastly, 15 August is a public holiday in France (as for most southern European countries) but not in Germany.

The bad luck suffered by Charlois still lasts. When the original photographic plate of Eros was searched for as final proof for his redemption, it could not be found. The whole collection of Charlois' plates had disappeared from the archives of the Nice Observatory. In spite of this, the *Dictionary of Minor Planet Names* includes, in the entry on Eros: 'Discovered by G. Witt at Berlin and independently discovered by A. Charlois at Nice.'

On 18 March 2004 a 30-metre NEA came even closer to our planet, well inside the orbit of the Moon and to within only 40,000 km from the Earth's surface. Very close indeed!

In general, an asteroid belongs to the NEA population if its minimum distance from the Sun is less than 1.3 AU. NEAs are further classified as Apollo, Amor and Aten asteroids, depending upon the size of the orbit compared to the Earth's perihelion (0.983 AU) and aphelion (1.017 AU) distances (Figure 6.5):

Apollos Named after asteroid (1862) Apollo. Semimajor axis larger than 1 AU (orbital period longer than 1 Earth-year), and perihelia less than 1.017 AU. Depending upon the orientation of their orbit in space they can cross the orbital path of the Earth.

Amors Named after asteroid (1221) Amor. Semimajor axis larger than 1 AU, and perihelia between 1.017 AU and 1.3 AU. They do not cross the orbital path of the Earth, and a collision with our planet is possible only if the asteroid's perihelion exactly matches the Earth's aphelion.

Atens Named after asteroid (2062). Semimajor axis less than 1 AU, and aphelia greater than 0.983 AU.

Asteroids with orbits entirely inside that of the Earth (with an aphelion smaller than the Earth's perihelion) have been only recently identified, and are usually referred to as IEAs (inner Earth asteroids). They are very difficult to discover because their position in the sky is always close to the Sun and therefore observations from the ground are disturbed by evening or morning twilight.

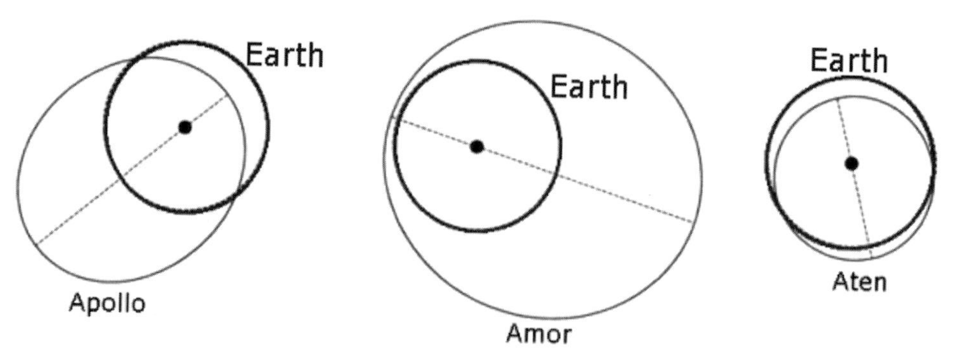

FIGURE 6.5. Typical Apollo, Amor and Aten NEA orbits.

It is believed that there are about 1 million NEAs with a diameter of the order of 10 metres, possibly up to 300,000 objects 100 metres in size, and 1,000 kilometre-sized asteroids, while only a few tens have dimensions larger than 10 km.

The long-term dynamical evolution of the NEAs is dominated by chaotic dynamics, and close encounters with the terrestrial planets are frequent. Because of this an asteroid is subjected, throughout its lifetime, to orbital changes large enough to allow transitions among the above-mentioned NEA types. As an example, an object presently classified as an Apollo could well have been an Amor in the past and will probably become an Aten at some time in the future.

When focusing on the risk of collision with the Earth, a subset of the NEA population is also identified. Potentially hazardous asteroids (PHAs) are objects larger than 100 metres which can in principle approach the Earth closer than 0.05 AU (7.5 million km – a little less than twenty times the Earth–Moon distance). The criterion for selecting PHAs is based on the computation of a critical quantity: the minimum orbital intersection distance (MOID), which monitors the behaviour in time of the least separation between the orbit of a NEA and that of the Earth. A small MOID does not necessarily mean that an actual close encounter with our planet will take place; such an event implies that both the asteroid and the Earth must be arriving at the same time at the positions along their orbits corresponding to the MOID. As seen in Chapter 3, resonances may provide dynamical sheltering from dangerous close approaches. Moreover, due to the chaotic nature of the motion of NEAs, some asteroids can repeatedly lose or gain PHA status, depending upon the perturbations affecting orbital evolution. However, continuous monitoring for PHAs is essential for determining the near-future collision risk of a cosmic impact on our planet. Table 6.4 shows some close encounters of PHAs with the Earth over the next 150 years.

For a better understanding of the data in Table 6.4 it is useful to compare them with two reference distances within near-Earth space: the distance of the Moon (about 380,000 km) and of the geostationary ring of telecommunication satellites (42,000 km).

Table 6.4. Encounters with potentially hazardous asteroids over the next 150 years.

Name	Date of encounter	Distance (km)
(99942) Apophis	2029 April 13	37,500
2000 WO107	2040 December 1	243,000
2001 WN5	2028 June 26	250,000
1998 OX4	2148 January 22	300,000
1999 AN10	2027 August 7	397,000
(35396) 1997 XF11	2136 October 28	413,000
2004 XP14	2006 July 3	432,500
1998 MZ	2116 November 27	436,500

Data from http://cfa-www.harvard.edu/iau/lists/PHACloseApp.html.

Due to the operation of automatic sky surveys using wide-field high-sensitivity telescopes, the number of known NEAs has quickly grown in recent years. In 1998 – a century after the discovery of Eros – their number was less than 1,000; but this figure had risen three times higher in 2005, and will probably grow even faster as new dedicated systems are realised.

Nevertheless, discovering NEAs is not the only critical business for assessing collision hazard. If discovery circumstances are unfavourable there could be insufficient data for computing a reliable orbit. Similar to what happened with Ceres (as discussed in Chapter 1), a NEA can become unobservable for quite a long time, to the point of being lost. The most striking example is that of (719) Albert – the second NEA to be found, and recovered nearly 89 years after its original discovery. Follow-up observations are essential for quickly obtaining good-quality orbital parameters, and to this end a large network of observatories – including those built by amateur astronomers – has been established. Astronomers cannot sleep while knowing that an 'out-of-control' and potentially hazardous object might be approaching the Earth.

IMPACT PROBABILITY

The collision of a NEA with the Earth can have widely different consequences from the fall of a meteorite to a mass extinction event – mostly depending on the size of the impactor. The collision of a 1-km object with the Earth – leading to dramatic climate changes on a global scale – is expected to occur over very long timescales. Impacts by 100-metre asteroids cause regional to global disasters, depending on the landing site. If the body hits the solid ground, then the formation of a crater a few kilometres wide is expected and the most severe consequences are therefore limited to the area of the impact. An asteroid smashing into the surface of an ocean will raise a tsunami, affecting continental coastlines on a worldwide scale. The threshold between a 'dangerous' object and a 'harmless' object is at a diameter of about 50 metres – the exact figure being dependent on its internal structure and composition. A monolithic metallic asteroid is less likely to disintegrate in the atmosphere than is a loosely bound object composed of rocks and ice.

Recalling the NEA size distribution, the larger the body, the less the probability of impacting the Earth (as shown in Table 6.5), which provides the impact frequency for different diameter ranges of the impactors.

The problem of dealing with statistics is that an event with a certain probability of occurring within a particular timespan could actually happen tomorrow or after a period of time longer than the statistical value. If there is no statistical meaning there is no certainty. As an example of how 'unlikely' a statistic can be, in 1971 the roof of a house in Wethersfield, Connecticut, was badly damaged by a meteorite weighting about 340 grammes. Eleven years later, another meteorite hit a rooftop in the same town! Computing the probability of such a 'twin event' would certainly result in a very small number. Yet it happened.

Table 6.5. Impact frequency as a function of the diameter d of the colliding body.

Diameter	Frequency (years)
$d>10$ km	50,000,000
1 km$<d<$10 km	500,000
100 m$<d<$1 km	5,000
30 m$<d<$100 m	500

The recent introduction of the Torino Impact Hazard Scale (Figure 6.6) has proven very useful. This scale is named after the Italian town of Turin, where in 1999 it was approved in its final form during a meeting dedicated to NEA impact hazard. The scale provides an estimate of the risks related to the collision of asteroids or comets with the Earth by using integer numbers ranking from 0 (probably no damage at all) to 10 (certainty of the occurrence of a global catastrophe). It represents a useful working tool for quickly assessing the level of hazard associated with the ever-increasing number of NEAs discovered. A short description of the risks implied by each level of alert is also added for communicating its exact meaning to the public at large.

No Hazard (White Zone)	0	The likelihood of a collision is zero, or is so low as to be effectively zero. Also applies to small objects such as meteors and bodies that burn up in the atmosphere as well as infrequent meteorite falls that rarely cause damage.
Normal (Green Zone)	1	A routine discovery in which a predicted pass near the Earth poses no unusual level of danger. Current calculations show that the chance of collision is extremely unlikely, with no cause for public attention or concern. New telescopic observations very likely will lead to re-assignment to Level 0.
Meriting Attention by Astronomers (Yellow Zone)	2	A discovery, which may become routine with expanded searches, of an object making a rather close but not highly unusual pass near the Earth. While meriting attention by astronomers, there is no cause for public attention or concern, as a collision is very unlikely. New telescopic observations will probably lead to reassignment to Level 0.
	3	A close encounter, meriting attention by astronomers. Current calculations predict a 1% or greater chance of collision capable of localised destruction. New telescopic observations will probably lead to reassignment to Level 0. Attention by public and by public officials is merited if the encounter is less than a decade away.

	4	A close encounter, meriting attention by astronomers. Current calculations give a 1% or greater chance of collision capable of regional devastation. Most likely, new telescopic observations will lead to re-assignment to Level 0. Attention by public and by public officials is merited if the encounter is less than a decade away.
Threatening (Orange Zone)	5	A close encounter posing a serious but still uncertain threat of regional devastation. Critical attention by astronomers is needed to determine conclusively whether or not a collision will occur. If the encounter is less than a decade away, governmental contingency planning may be warranted.
	6	A close encounter with a large object posing a serious but still uncertain threat of a global catastrophe. Critical attention by astronomers is needed to determine conclusively whether or not a collision will occur. If the encounter is less than three decades away, governmental contingency planning may be warranted.
	7	A very close encounter with a large object, which if occurring this century, poses an unprecedented but still uncertain threat of a global catastrophe. For such a threat in this century, international contingency planning is warranted, especially to determine urgently and conclusively whether or not a collision will occur.
Certain Collisions (Red Zone)	8	A collision is certain, capable of causing localised destruction for an impact over land or possibly a tsunami if close offshore. Such events occur on average between once per 50 years and once per several thousand years.
	9	A collision is certain, capable of causing unprecedented regional devastation for a land impact or the threat of a major tsunami for an ocean impact. Such events occur on average between once per 10,000 years and once per 100,000 years.
	10	A collision is certain, capable of causing a global climatic catastrophe that may threaten the future of civilisation as we know it, whether impacting land or ocean. Such events occur on average once per 100,000 years, or less often.

FIGURE 6.6. The Torino Impact Hazard Scale. (http://neo.jpl.nasa.gov/torino_scale.html.)

Apophis: a Christmas gift

On Christmas Day 2004 the mobile telephones of astronomers all over the world began ringing furiously. The 'hot news' was that an asteroid – provisionally named 2004 MN4, and some 300 metres in diameter – was dangerously climbing the Torino Scale. The events happening before and after that day are a good example of what monitoring a potentially hazardous asteroid actually means.

2004 MN4 was discovered on 19 June 2004, and was again observed on 18 December. Orbit computations showed that there was a non-negligible impact probability with the Earth at some time in the future, corresponding to a value of 2 on the 10-point Torino Scale. It was not an unusual result, and not particularly alarming. The large uncertainties associated with determining an orbit from a few observations often produce similar situations, because of the many trajectories compatible with the data. As new and more precise observations become available and the range of possible trajectories shrinks, the collision solution is usually ruled out and the Torino ranking drops to a safe 0. But this was not the case for 2004 MN4. In spite of improving knowledge of its orbit, the possibility of a collision with our planet on 13 April 2029 remained, until reaching a value of 1 over 40, corresponding to level 4 on the Torino Scale. It had never happened before, and by Christmas Day new and more precise observations were urgently required from around the world.

An archival search for pre-discovery observations (see Chapter 1) was therefore initiated, and on 27 December the object was found on an image dated 15 March 2004. The observed orbital arc extended significantly, and the consequent sharp improvement to the orbit solution allowed any chance of an Earth impact on 13 April 2029 to be ruled out. Collision was replaced by an extremely close encounter with our planet, when the asteroid will be so bright that it will be visible by the naked eye in Europe, Africa and part of Asia (Figure 6.7 and Table 6.4).

Asteroid 2004 MN4 is now catalogued as number 99942, and bears the name Apophis, after the Egyptian god Apep – 'The Destroyer'.

DON'T PANIC!

As the number of PHAs increases, the attention of the media and of the public at large increases accordingly, often resulting in alarm that is not entirely justified. On the other hand it is not easy for scientists to communicate the complex interplay of dynamical considerations underlying the evaluation of a cosmic impact hazard.

When browsing one of the many astronomical calendars available on the Internet, side by side with classical celestial phenomena such as eclipses and occultations, an ever-increasing number of close encounters of asteroids with the Earth is found. But - don't panic - this does not mean that an increasing number of NEAs are actually passing close to our planet. That would surely be a source of

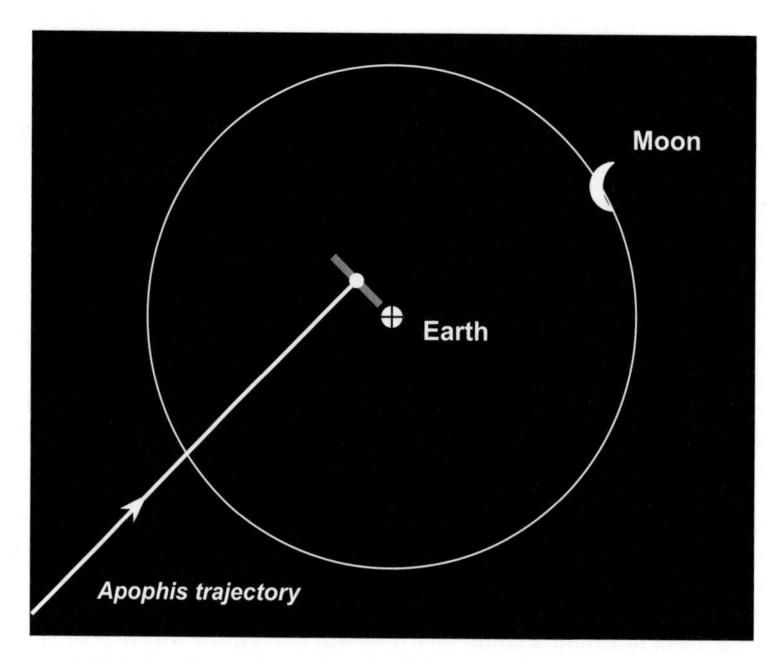

FIGURE 6.7. The 2029 encounter of Apophis with the Earth. The line perpendicular to the trajectory of the asteroid represents the error with which the position of the asteroid is known. It does not touch the Earth, and an impact can therefore be safely ruled out.

deep concern. Instead, it is good news instead, as the increasing number of close encounters on record reflects the improvement in our ability to detect smaller and smaller NEAs as they routinely approach our planet. In other words, we are beginning to fill the observational gap between asteroids and meteorites. No wonder, then, that 'minimum-distance' record-breaking events often happen and are presented as 'breaking news', thus mistakenly giving the impression of a growing hazard. Then, as we have seen in the case of Apophis, the risk suddenly disappears, leading to a feeling of unjustified alarm.

Celestial mechanics is largely responsible for this. In general, when a potentially dangerous event is identified, it is common that the probability decreases steadily as appropriate safety actions are undertaken, until its value becomes small enough to be safely neglected. Dealing instead with orbital motion and with the uncertainties associated with orbit determination, the probability of collision exhibits counterintuitive behaviour by either growing or dropping to zero (as illustrated in Figure 6.8). Moreover, it should always be kept in mind that an object has to be towards the top of the Torino Scale before a collision can be considered as an almost certain event. A careful reading of the explanation of level 4 – the highest score to date – is recommended.

Yet 'don't panic' does not necessarily mean 'don't worry'. There are two highly sophisticated systems for continuous monitoring of the entire NEA population. One of them is located at the Jet Propulsion Laboratory as part of the

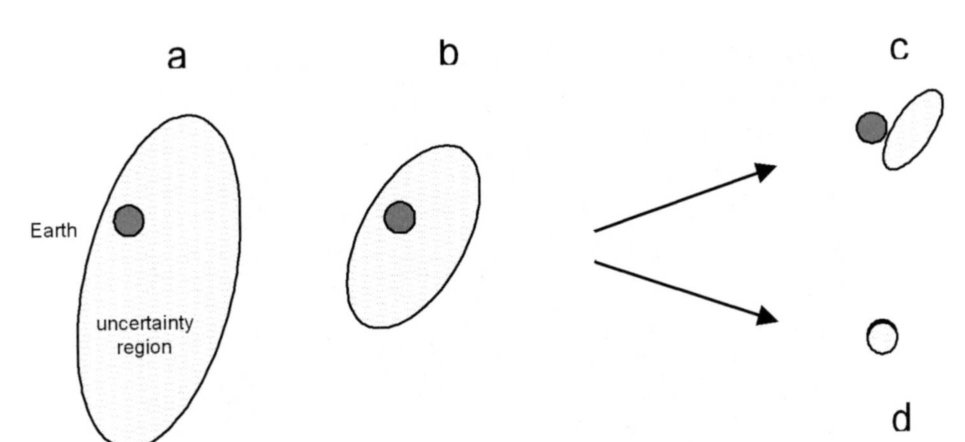

FIGURE 6.8. (a) Impact probability can be graphically represented by the ratio between the width of the circular section exposed by our planet to an incoming celestial body and that of the uncertainty region which encompasses all possible orbital paths of the impactor. (b) As the orbit of the impactor is determined with increasing accuracy the uncertainty region shrinks. If the Earth is still present inside, then the impact probability grows. There are only two possible results of this process: either the uncertainty region, while shrinking, leaves our planet outside (c) (and then the probability drops to zero); or the uncertainty region coincides with or is inside the figure of the Earth (d) (a collision is therefore unavoidable).

NASA Near-Earth Objects Programme, and the other (NEODys) has been developed at the University of Pisa. They allow a real-time assessment of the collision probability of newly discovered objects, as well as performing periodic updating of our knowledge of the dynamical and physical properties of all known NEAs.

Every time a NEA is discovered and the observational data are available online, an automatic procedure is started to perform all sorts of celestial mechanics computations, including the identification of possible dangerous near-future encounters with the Earth. If the latter is the case, a warning message is sent to the astronomical community for focusing the observational efforts in order to improve the accuracy of the orbit determination of the potentially dangerous asteroid. The whole procedure is then repeated until a collision with our planet can be safely ruled out.

The case of 2004 MN4 Apophis can be considered as a successful test for proving the efficiency of these early warning systems. They also allow us to understand the three major goals of current NEA research:

- To improve surveys for discovery and follow-up observations in order to increase the number of NEAs known and to avoid the observational loss of a potentially hazardous object.
- To increase ground-based observations aimed at determining the compositional parameters of individual asteroids.

- The realisation of *in situ* exploration by means of dedicated space missions.

These operations will soon provide the ground truth needed for planning and implementing efficient safety strategies.

MITIGATION

Public institutions have recently faced the problem of the hazard posed by NEAs, including cosmic impacts among possible natural disasters. A resolution of the Council of Europe issued on March 1996 states: 'Although, statistically speaking, the risk of major impacts in the near future is low, the possible consequences are so vast that every reasonable effort should be encouraged in order to minimise them'.

Under these auspices was instituted the Spaceguard Foundation – a non-profit organisation aimed at coordinating both theoretical research and astronomical observations for protecting our planet from cosmic hazards.

Familiar natural disasters – such as volcanism and earthquakes – cannot be completely avoided, but their impact on human activity is mitigated. Houses are built according to seismic criteria, and the direction of lava flows is changed in order to protect human settlements. But the case for NEAs is peculiar, because mitigation strategies foresee the possibility of a complete removal of the hazard by changing the trajectory of the impactor by the exact amount so that it will miss the Earth. Deflecting an asteroid *en route* to a collision with our planet can be achieved by different means:

- Sending nuclear missiles toward the impactor in order to break it into pieces and/or deflect its trajectory.
- Landing on the impactor and installing, on its surface, an engine – a 'cosmic tugboat' – capable of gently pushing the asteroid onto a safer trajectory.
- Heating the asteroid's surface with a powerful laser, to the point of exploiting the propulsion provided by the onset of natural jets of gas and dust.
- Changing the reflecting properties of the impactor (e.g. by covering a large fraction of its surface with mirrors), in order to strengthen certain orbital perturbations such as Yarkowsky drift and solar radiation pressure.
- Impacting the asteroid with a 'suicide' spacecraft as massive as possible and at the highest possible relative velocity, in order to provide an impulse large enough to significantly deflect its trajectory.

The efficiency of the different deflection techniques depends upon the individual case, being strongly affected by the physical properties of the impactor. A 'rubble pile' asteroid (a loosely bound aggregate of fragments reaccreted after a major fragmentation event) could be easily destroyed, while a solid rock or metallic monolite would result in a completely different deflection scenario. Outgassing can be triggered only for rocky–icy bodies, and needs careful modelling. Nuclear weapons should be used only in case of emergency, as, for example, when the impactor is discovered rather late and there is no time for planning alternative strategies. Fragmentation is not always considered a good option, because it could result in an increase in the number of impactors; and changing the physical properties of the asteroid requires sophisticated and complex mission profiles.

The mitigation strategy that appears conceptually simple, highly effective and technically feasible is that of changing the asteroid's motion by an 'artificial cratering event' (the last option in the above list). It has been demonstrated that the chaotic nature of a NEA's orbital motion can be exploited for transforming a small impulse into a large deflection. The only constraint is that to be effective the small impulse should be applied as soon as possible: the earlier the deflection (with respect to the impact date), the smaller the impulse, and the easier it would be to apply it from a technological point of view.

An even more favourable situation occurs if the orbit of the asteroid is in a mean motion resonance with that of the Earth, because in this case there could be repeated close encounters with our planet prior to impact. These *resonant returns* in the neighbourhood of the Earth can be exploited for further amplifying

Don Quijote

One of the problems of the NEA population is their diversity: small monoliths, remnants of metallic cores, and rocky and icy irregular bodies carrying the traces of an intense collisional evolution are mixed up within NEAs. Yet determining the physical properties of NEAs is essential for planning successful mitigation strategies; therefore several missions have been realised or are being planned for selected NEA targets. In 2000 the NEAR (Near Earth Asteroid Rendezvous) spacecraft reached Eros and orbited around it for more than a year. It then soft-landed on the asteroid, after which its mission was terminated. The current Japanese Hayabusa mission, after having successfully reached asteroid Itokawa in 2005, will hopefully return the first asteroidal samples to Earth. In July 2005 the spectacular collision of the Deep Impact spacecraft with comet Tempel 1 demonstrated our ability to hit a small body travelling at high relative velocity. The next logical step is a precursor mission for the full testing of an asteroid deflection strategy, and a sophisticated European project could be soon fulfilling this need. The mission profile envisages two spacecraft – one of which will orbit the target asteroid, while the other will be sent on a collision trajectory. The role of the orbiter is twofold: monitoring the impact and allowing precise determination of the orbit of the asteroid before and after the event in order to measure the expected deflection. This mission is named Don Quijote, after the celebrated work by Miguel Cervantes, and denoting the noble intent of the two characters who give their names to the spacecraft: the impactor Hidalgo and the orbiter Sancho.

the effect of a deflection. This is exactly what is being studied for Apophis. Ruling out the 2029 impact did not render Apophis less dangerous. When its near-future orbital evolution was investigated, other possible collisions with the Earth were revealed – the earliest occurring on 13 April 2036. Far from creating unjustified alarm, Apophis represents the perfect case for proving our ability to save our planet from a cosmic impact. Hopefully there is plenty of time for developing the necessary technology and for realising space missions dedicated to testing mitigation by deflection... just in case of panic.

7

Of Moon and man

Che fai tu, Luna in ciel? Dimmi, che fai silenziosa Luna.
What do you do, Moon, aloft? Let me know, what, silent Moon, you do.

Giacomo Leopardi, 'Night song of a wandering Asian shepherd'

After reaching the climax of the historical space race between the Soviet Union and the United States, the popularity of the Moon has been seriously compromised by thirty years of Solar System exploration. The Voyagers' 'grand tour' of the outer planets unveiled the secrets of complex ring systems and of immense cloudy atmospheres, but above all the twin spacecraft sent back to Earth spectacular images of a multitude of brand new planetary moons. The Galileian satellites – different from each other as much as they are dynamically tightly bound – are four new worlds to explore. The mysterious atmosphere of Titan has for two decades fascinated astronomers, until in January 2005 ESA's Huygens probe pierced the layers of dense clouds and landed on a frozen methane landscape. The scars on the surface of Miranda... the geysers spouting from the interior of Triton... the battered surface of Mimas... the bright white globe of Enceladus... to name a few, have attracted the interest of scientists and of the public at large. However, the Moon – *our* Moon – remains a celestial body unique in the Solar System. The lunar motion has puzzled astronomers of all epochs, and its influence on life and widespread religions and cults has followed the development of man and civilisation. The return of humans to the surface of the Moon has been recently announced – this time, to stay.

THE CYCLES OF SELENE

In many respects we live in a binary system – a double planet (Figure 7.1). Indeed, the mass of the Moon amounts to a significant fraction of that of the Earth (1:81), thus representing a major anomaly with respect to the other planet–satellite pairs in the Solar System (the second largest being Triton/Neptune = 1/740). The distance of the Moon is large enough to be subjected to consistent solar perturbations, which result in periodic and secular variations of its orbital parameters (Figure 7.1). Semimajor axis, eccentricity and inclination

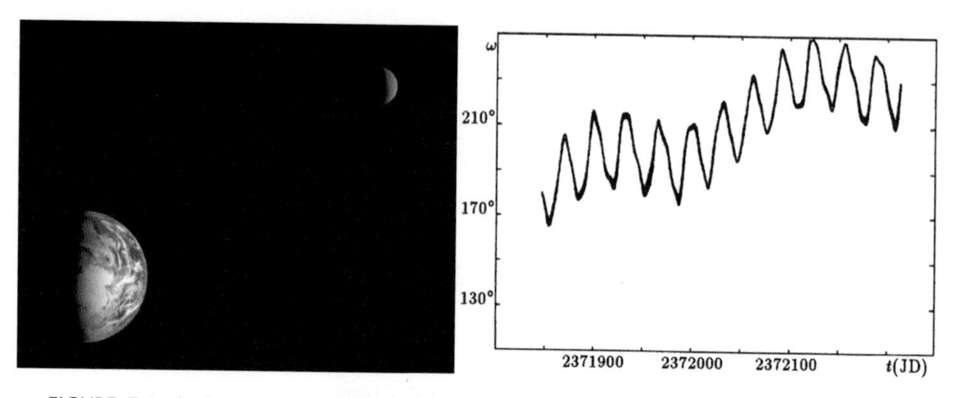

FIGURE 7.1. (Left) The Earth–Moon system imaged by the NEAR spacecraft during its south-polar Earth gravity assist on the way to asteroid Eros. (Courtesy NASA.) (Right) The oscillations of the lunar perigee are of considerable amplitude and follow long-period cycles known as *evection*.

exhibit complex patterns, while the nodal and apsidal lines follow precessional (the former) and prograde (the latter) motions. Finally, the orbital plane of the Moon is not equatorial, but is inclined at about 5° to the ecliptic.

Although the regular appearance of the lunar phases was possibly the first celestial calendar used by mankind (the ancient Jews used to say: 'The Moon has been created to number the days'), understanding in detail the motion of our satellite is not an easy task. Lunar dynamics belongs to the general three-body problem, which is well known to be non-integrable, and computing accurate ephemerides of the Moon has been historically challenging for celestial mechanics. It is reported that Isaac Newton found the lunar problem so difficult that it made his head ache and kept him awake so often that he would think of it no more.

Yet the many signs of the presence of the Moon in our life are mostly connected with the periodicities that can be found in its orbital motion. Lunar phenomenology has been extensively studied since ancient times, leading to the discovery of the *lunar cycles* – numerological rules which allowed the prediction, with remarkable accuracy, of astronomical events such as phases and eclipses. As already pointed out when discussing Bode's law or the overabundance of resonances in the Solar System, the existence of peculiar numeric relationships can be considered a lucky chance. In order to verify this luck, celestial mechanics tries to understand if and how a dynamical explanation can be found. The most important lunar cycles are listed in Table 7.1.

In ancient times, several discoveries concerning lunar dynamics depended upon specific events, such as the almost exact repetition of easily observable dynamical configurations. The first cycle of Table 7.1 is named after the Greek astronomer Meton (*c.*440 BC) and chiefly involved in the compilation of calendars.

It is common experience that if the phase of the Moon is observed at a certain date and is checked on the same day a year later, our satellite exhibits a different

Table 7.1. Lunar cycles.

Cycle	Duration (years)	P	C	E	d_M	d_S
Metonic	19.00	x	x	–	–	x
Saros	18.03	x	–	x	x	x
Hipparchus short	20.29	x	–	–	x	–
Hipparchus half-long	345.00	x	x	–	x	x
Hipparchus medium	441.29	–	–	x	–	–
Hipparchus mega	689.99	x	–	x	x	x

x Can be used for predicting the following:
P Lunar phases
C Calendar dates
E Eclipses
d_M Apparent diameter of the Moon at eclipse
d_S Apparent diameter of the Sun at eclipse

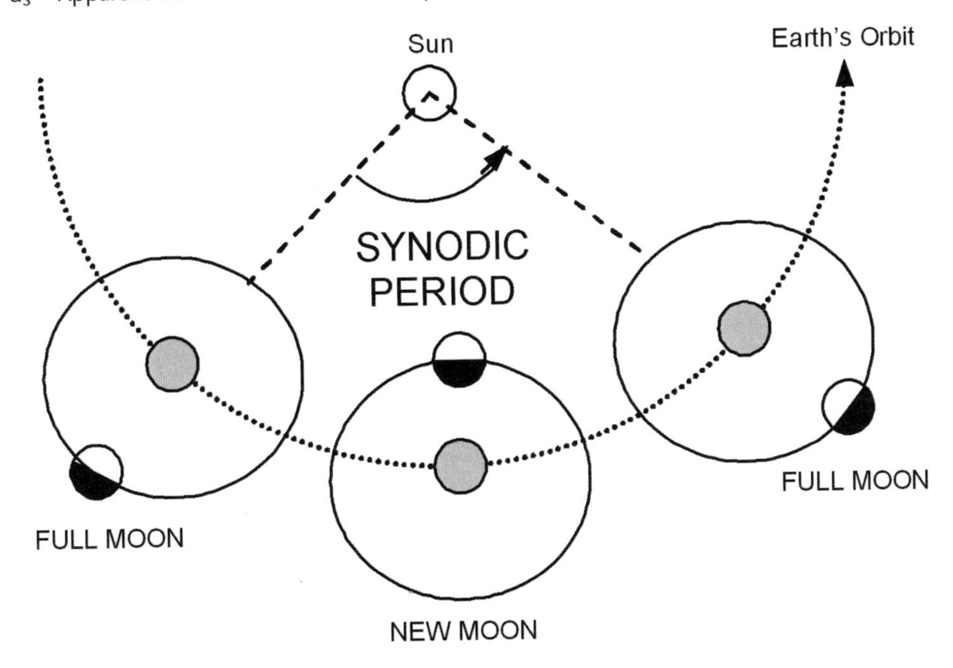

FIGURE 7.2. A conjunction occurs when the Moon reaches the minimum distance from the Sun and corresponds to the observable new Moon. At opposition the distance from the Sun is at maximum and a full Moon is observed. The synodic period is the time elapsed between two identical lunar phases; for example, two subsequent full Moons. In the diagram the inclination of the lunar orbital plane has been discarded.

phase. This simple experiment demonstrates that in one year the Moon does not complete an integer number of phase cycles. Quantifying this phenomenon is to define the *synodic period* of the Moon – the time elapsed between two subsequent conjunctions with the Sun (Figure 7.2), thus encompassing a whole phase cycle. Its duration is therefore easy to compute: 29.53 days.

Using the lunar synodic period as the length of a month (it is often referred to as a 'lunar month') would be extremely useful, because the Moon, while passing from waxing to waning, acts as a huge celestial clock. Unfortunately it is also an impractical choice, because 29.53 × 12 = 354.36 days, so that the 11 days left to complete the 'astronomical year' (the time that Earth takes to complete an orbit) would soon accumulate, with annoying consequences. In less than 20 years Christmas would be celebrated during the summer in the northern hemisphere! How is it possible, then, to harmonise the motion of the Earth with that of the Moon?

Meton discovered a connection between the year and the lunar phases. He observed that the number of days contained in 19 years is very close to the number of days contained in 235 synodic periods of the Moon. For modern celestial mechanics Meton discovered a commensurability between the orbital motions of the Earth and of the Moon:

$$19 \times 365.24 = 6{,}939.6 \text{ days}$$
$$235 \times 29.53 = 6{,}939.7 \text{ days}$$

The decimal figures appearing in the length of the year account for the leap years.

There is no dynamics underlying Meton's cycle. In ancient times it simplified the compilation of solar and lunar calendars, which ruled many aspects of civil and religious life (for example, the computation of the date of Easter). Today the cycle is no longer required, but there is an interesting observable consequence. If you look at sky on the day of your 19th, 38th, 56th... (all multiples of number 19) birthday, the Moon appears in the same phase as when you were born!

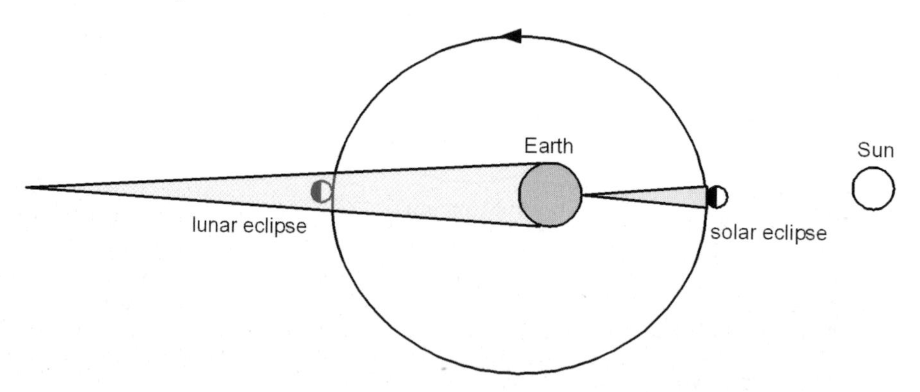

FIGURE 7.3. The different size of the Earth and Moon implies that at solar eclipse only a given region on the surface of our planet is interested by the phenomenon, while at lunar eclipse the Moon is completely obscured by the Earth's shadow, and the event is observable from any place where the Moon is above the horizon.

Next in Table 7.1 is the Saros period of 18 years 10 days (or 11 days, depending on the number of leap years within the cycle), which provides a simple algorithm for eclipse prediction. In trying to improve the predictability of events in the Earth–Moon–Sun system, Hipparchus (*c*.140 BC) introduced four additional relationships. Nevertheless, as the period of a cycle increases, its applicability becomes more and more impractical.

ACTS OF THE GODS

An example of the intrinsic complexity of lunar motion is provided by the eclipse prediction cycle known as the Saros; but when we try to produce a modern interpretation of its existence, we become caught between old and new celestial mechanics, traditional orbit computation and digital computers, archeoastronomy and chaos.

Now that accurate lunar ephemerides can be easily found on the Internet, or generated using a personal computer, the occurrence of solar and lunar eclipses is well known in advance, representing a fascinating astronomical event. But in ancient times it was not so. Eclipses were of minor scientific interest, but were of great political and social significance. The following is the report of a Babylonian–Assyrian astrologer, who lived in Mesopotamia – the region of the Middle East between the rivers Tigris and Euphrates – around 650 BC: 'On the 14th an eclipse will take place; it is evil for Elam and Amurru, lucky for the king, my lord; let the king, my lord, rest happy. It will be seen without Venus. To the king, my lord, I say: there will be an eclipse. From Irasshi-ilu seniore, the king's servant.' (R.C. Thompson, *The Reports of the Magicians and Astrologers of Ninevah and Babylon*, 1900, p.273.)

Among the usual good auspices to the king, the almost dramatic way in which the imminent eclipse is repeatedly stressed emphasises its social importance and thus the need for the king to know of it in advance. The sudden darkening of the Sun (Figure 7.4) and the appearance of stars during daytime, which characterises a total solar eclipse, or the reddening of the full Moon at night, must have been frightening as evil omens from the gods. But how could Irasshi-ilu

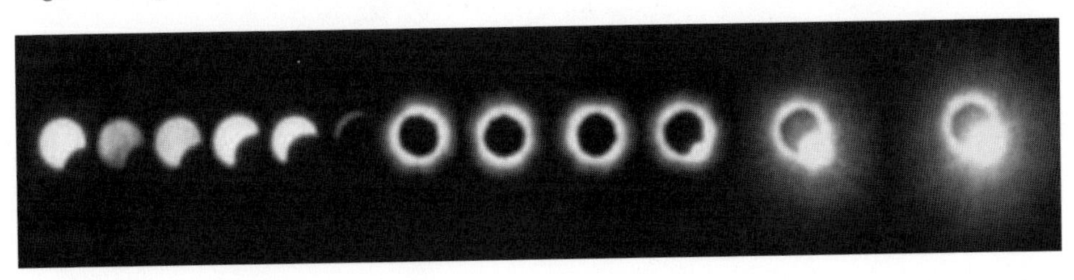

FIGURE 7.4. Solar eclipses are effective because the apparent diameters of the Moon and the Sun are almost the same (about 0°.5). This is due to the peculiar coincidence that the ratio of the true diameters of Sun and Moon is approximately the same as the ratio of the distances of the Earth from the Sun and from the Moon respectively.

predict the occurrence of an eclipse almost 3,000 years ago? Even for celestial mechanics it is not a trivial task.

Eclipses occur when rather peculiar dynamical configurations are satisfied, with Earth, Moon and Sun aligned to a high degree of precision (Figure 7.5). If the orbit of the Moon around the Earth and that of the Earth around the Sun were lying on the same plane, monthly lunar and solar eclipses would be assured. As already discussed in the previous section, the Moon's orbit is inclined with respect to the ecliptic and its motion is strongly perturbed by the Sun. In particular, the behaviour of the line of nodes (the intersection between the two orbital planes (Figure 7.5)) follows a complex pattern, so that it is quite difficult

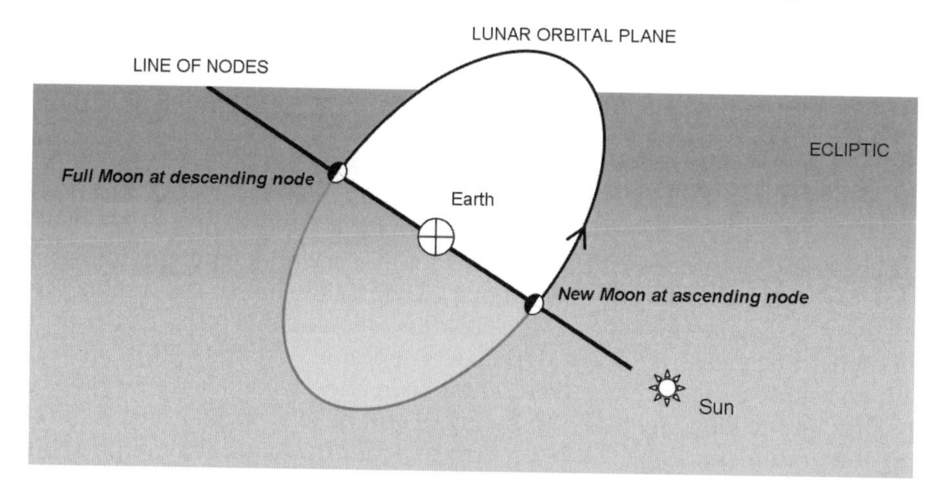

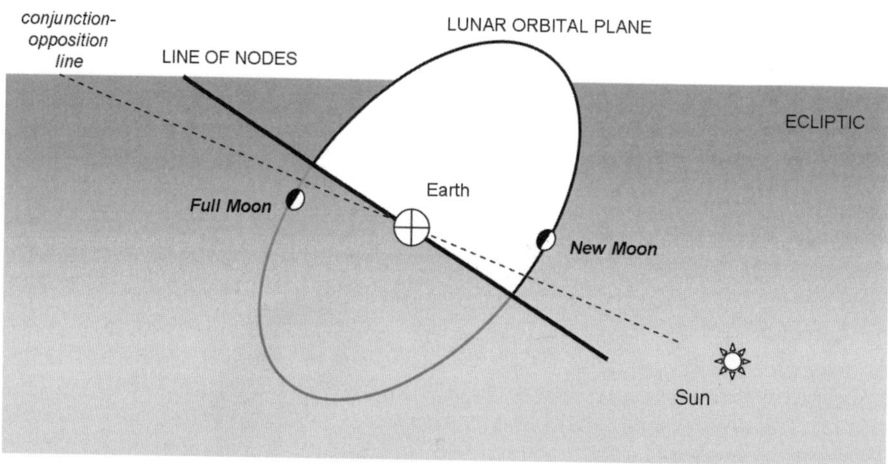

FIGURE 7.5. Solar and lunar eclipses occur when the conjunction–opposition line matches the line of the nodes at new Moon or at full Moon respectively. The upper diagram shows the geometries of solar and lunar eclipses, while the lower diagram shows why conjunctions and oppositions can occur without satisfying eclipse conditions.

to predict when it becomes aligned with the Sun–Earth direction – a necessary condition for an eclipse to occur. Predicting the dynamical evolution of the three-body problem, Sun–Earth–Moon, was certainly not possible in Babylonian times, and Irasshi-ilu could perform only very simple calculations.

The answer must be sought in the huge archives of cuneiform writings kept by the ancient Babylonians. Thousands of tablets have been found during archeological excavations, and many of them report astronomical observations extending back over many centuries. Although Mesopotamia was not the best location for observing the sky, due to the hot desert climate and frequent dust-storms, Babylonian astrologers were the most constant observers of celestial phenomena in antiquity. In addition, they duly recorded their observations in archives, and it was probably by looking at this long record of observations that they discovered that lunar eclipses did not take place at random, but in series of five or six eclipses separated by longer time intervals. The Babylonians relied mainly on lunar eclipses because the Earth's shadow is much larger than the Moon's – big enough for the whole Moon to be immersed in the shadow – and when a lunar eclipse occurs, it is visible from half of the Earth at any time. By contrast, the narrow track swept by the Moon's shadow during a total or annular solar eclipse is, at most, only a few hundred kilometres across, although a partial eclipse will be visible over a considerably larger area, but still much less than half the Earth's surface (Figure 7.3).

The Babylonian empire ruled the region from about 2000 to 700 BC, after which the Assyrians – a military people from the northern regions of Mesopotamia – displaced them. The Assyrians assimilated much of the Babylonian culture, and continued to observe the heavens to read the signs traced by the gods in the sky. After further political turmoil, the Persians

The price of ignorance

Today eclipses are not frightening, and we can fully enjoy their beauty; but many historical reports have reached us concerning the peculiar events which accompanied their occurrence. In China, during a solar eclipse the imperial guard was told to play drums as loud as possible in order to set the Sun free; and from the same country comes the sad story of the two astronomers Hi and Ho, who were executed for failing to predict an eclipse. Knowledge of the Saros could have easily saved their lives.

Much later, and in another part of the world, Columbus exploited a lunar eclipse for his own purposes. During his fourth journey to America he found himself without food and water, while surrounded by hostile natives. He therefore threatened the Indians, pretending that he had the power to obscure the Moon if they did not surrender and provide supplies. In reality Columbus knew that a lunar eclipse was about to take place on 29 February 1504, as reported in the lunar ephemerides. The Indians were deeply frightened, and approached Columbus's ship in submission. He appeared only at the end of the eclipse, as if he could control celestial phenomena.

conquered Babylon toward the end of the sixth century BC. It was probably around this time that the Chaldeans – who also lived in lower Mesopotamia – refined the ancient Babylonian eclipse observations and discovered the Saros.

'Saros' means 'repetition', and indicates a period of 6,585.277 days (18 years 10/11 days. dependent on the number of leap-years within the cycle) after which a given sequence of lunar and solar eclipses repeats. Eclipse prediction then becomes a very simple matter of adding calendar dates. As an example, taking as a reference the solar eclipse of 11 August 1999, it is possible for anyone to compute that a solar eclipse had already occurred on 31 July 1981 and will be happening again on 21 August 2017, before and after 18 years 10 days – one Saros.

ECLIPSED BY THE SAROS

Correct computation and prediction of natural phenomena is one of the major goals of science, and the discovery of the Saros was a great scientific achievement. In this respect the keyword is 'periodicity': the more periodic an event, the more accurate its prediction. Periodicity is, in turn, a keyword for modern celestial mechanics because, following Poincaré's words, it allows a deeper understanding of the dynamics of N-body systems.

The first step for investigating the complex interplay among gravitational perturbations which simplified the difficult task of eclipse prediction is therefore in providing a dynamical representation of the Sun–Earth–Moon system. A simplified lunar problem is obtained taking into consideration a 'double' two-body problem consisting of the unperturbed orbits of the Moon around the Earth and of the Earth around the Sun. During a complete revolution around our planet the Moon reaches a minimum (at conjunction) and a maximum (at opposition) distance from the Sun (Figure 7.2). Conjunction and opposition, as seen from the surface of the Earth, correspond to new Moon and full Moon respectively, because of the different illumination conditions as well as the offset of our satellite with respect to the ecliptic due to its orbital inclination. Solar and lunar eclipses occur only if the new Moon or the full Moon appear when our satellite crosses the ecliptic, thus being aligned in the Sun–Earth direction. In terms of orbital parameters, the Moon must be at one of its nodes while the line of nodes is aligned in the Sun–Earth direction (Figure 7.5).

The existence of the Saros implies that there must be a periodic repetition of certain peculiar geometries, showing that a resonance is acting within the system. In order to understand which commensurabilities are involved, one should consider that the period of revolution of the Moon around the Earth – the length of a lunar month – has many definitions. Indeed, on one side, while the Moon revolves around the Earth our planet moves along its orbit at a significant angular velocity (about 1° per day), and on the other side, the perturbations of the lunar orbit are strong enough to change its orientation in space on a rather short timescale. The lunar period therefore depends upon the reference frame adopted, leading to the following definitions:

- The *synodic month*, $T_S = 29.5306$ days, is the time between two subsequent conjunctions, and is therefore ruled by lunar phases.
- The *anomalistic month*, $T_A = 27.5545$ days, is the time between two passages of the Moon at perigee.
- The *nodal month*, $T_N = 27.2122$, is the time required for the Moon to return to the ascending node (crossing the ecliptic from south to north).

The last two values are different, as the apsidal line of the lunar orbit (the line joining perigee and apogee) is not fixed in space but rotates in its own orbital plane, making a complete turn in about 9 years, while the line of nodes spans a full 360° in 18.5 years.

Multiplying the values of the synodic, anomalistic and nodal months for suitable integer numbers, remarkably similar results are obtained:

$$223\ T_S = 6{,}585.32 \text{ days}$$
$$239\ T_A = 6{,}585.53 \text{ days}$$
$$242\ T_N = 6{,}585.35 \text{ days}$$

This commensurability relationship among the three different lunar months provides a modern definition of the Saros and justifies its existence. Suppose that a solar eclipse occurs at a given conjunction when the Moon is both at perigee and crossing the ascending node. Because of the different values of the lunar months, at the next conjunction (29.5306 days later) the Moon has already passed both perigee and the ascending node a couple of days before. Therefore no eclipse is likely to occur, since the Moon is not aligned with the Sun–Earth line, but is slightly tilted. But at the 223rd conjunction after eclipse, the Moon is also at its 242nd nodal passage and its 239th perigee, thus again fulfilling solar eclipse geometry.

The Saros is a powerful cycle, because it allows the prediction of precise timing (the slightly different figures involved in the commensurability translates into an error of about an hour after 18 years) and the observable characteristics of the eclipse (see Table 7.1). The whole solar disc is covered by the Moon due to a remarkable coincidence of the apparent diameters of the Moon and Sun. The former is smaller but relatively close to Earth, while the latter is a great deal larger and much more distant. If a solar eclipse occurs at apogee, the Moon is at its maximum distance from the Earth, and therefore does not completely obscure the Sun. This type of eclipse is called 'annular', because the bright edges of the Sun surrounding the Moon appear like a ring. A similar reasoning applies for the duration of a lunar eclipse, when the Moon passes the Earth on the opposite side from the Sun and is obscured by the Earth's shadow.

LUNAR THEORIES

Eclipses are easily observable naked-eye events, and it may be wondered if the existence of a repetitious cycle also implies a deep similarity between the whole orbital paths followed by the Moon from Saros to Saros. Imagine 'cutting' two subsequent 1-Saros-long lunar trajectories and superposing them. Will they almost perfectly match all the way through, or only near eclipses? This is the equivalent of asking whether the Moon is close to a periodic orbit with a period equal to one Saros.

An orbit of this type will be a winding open path which after 223 turns around the Earth (corresponding to 18 years 10/11 days) begins to follow its own footsteps. If this is so, an ellipse is no longer representative of the 'true' orbit of

The Delaunay orrery

The life and work of Charles Delaunay (Figure 7.6) progressed in parallel with that of another famous astronomer, Jean Joseph Urbain Leverrier, who discovered Neptune. A collision became inevitable when in 1870 Leverrier – who was extremely unpopular because of his antagonistic attitude toward his colleagues – was eventually removed from his position as Director of the Paris Observatory, to be replaced by Delaunay. These were rather tough political times. The French had been defeated by the German army, and the uprising in Paris had led to the establishment of the Commune. Delaunay spent much of his time trying to protect the observatory – but not for long. In 1872 he drowned, together with friends, in a boat accident during a trip at sea, and Leverrier was restored to his position as Director.

About a century after these events, Delaunay's theory was checked by using one of the first digital algebraic manipulators. Only about 20 hours of computer time were needed to reproduce Delaunay's twenty years of work; but much to his credit only two minor errors were found, and his theory is now fully validated.

FIGURE 7.6. Charles Delaunay and the 'grand coupole' of the Paris Observatory at Meudon.

the Moon, thus requiring a novel dynamical approach and mathematical formulation. This was probably why Newton suffered from headaches, because the lunar theories developed specifically for the motion of the Moon are among the most refined and complex applications of perturbation theories to date. The French astronomer Charles Delaunay (1816–1872) devoted twenty years of his life to the development of an analytical lunar theory, which still represents a milestone of celestial mechanics. The Hill–Brown theory, proposed at the beginning of the twentieth century, contains equations with more than 1,500 terms for describing in detail the perturbations acting on our satellite.

A numerical approach for finding Saros-like periodic orbits has recently been proposed by Giovanni B. Valsecchi and collaborators. Eclipses occurring with the Moon at perigee or at apogee can be considered as a practical example of the occurrence of mirror configurations (as defined in Chapter 2) in real systems. Starting from this consideration and relying on the importance of mirror configurations in the study of the N-body problem, Saros-like periodic orbits have been numerically found in the Earth–Moon–Sun system. Each one is characterised by commensurabilities among the synodic, the anomalistic and the nodical months. As already mentioned, the use of numerical methods produces less information on the perturbations acting within the system, but provides the possibility of finding periodic orbits closely resembling that of the Moon.

Figure 7.8 shows the behaviour of the inclination of the Moon for three different orbits. The middle curve is the 'real' Moon, the upper one is that of a periodic orbit having the length of the Saros (223 T_S = 6,585.32 days), and the lower refers to a periodic orbit with a much longer period of 1,751 T_S = 51,708.08 days (about 141.5 years). As predicted by Poincaré's famous conjecture (discussed in Chapter 2), the behaviour of both periodic orbits resembles the real orbit of the Moon, but the orbit with a longer period more closely matches it.

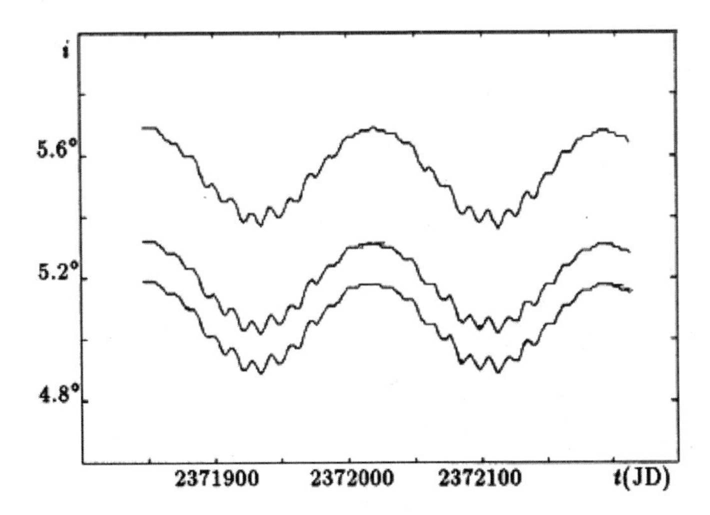

FIGURE 7.7. Perturbation of the inclination of the Moon for three different sample orbits.

FIGURE 7.8. The web of periodic orbits surrounding the Moon, the position of which is indicated by a full white circle.

A systematic search for longer and longer periodic orbits has led to the diagram in Figure 7.8, in which every dot represents a Saros-like periodic orbit. Far from being a rigorous demonstration of Poincaré's conjecture, the diagram shows it in action, demonstrating the dense population of periodic orbits in Earth–Moon space. Moreover, periodic orbits are not distributed at random, but follow a well-defined structure recalling the road map of an imaginary 'Sarostown'. A narrow web of smaller and smaller streets and alleys departs from the large boulevards, the intersection of which opens wide squares. The 'real' Moon is placed just at the bordering of the largest square in town, and again it might be wondered whether this is due to chance or has an underlying dynamical meaning.

HOLIDAYS IN ELATINA

We know that the tidal interaction between the Earth and the Moon (Chapter 1) slows the Earth's rotation while pushing the Moon away from our planet. If the distance of the Moon increases, its orbital velocity decreases and therefore the length of the various lunar months changes according to Kepler's laws. As a consequence, the present Saros is doomed to disappear. This raises a number of questions. How long has the 18-year eclipse prediction cycle been in existence? How long will it last? Is the Moon passing quickly from one Saros-like commensurability to the other, and why? Does the Saros have something to do with the long-term stability of lunar motion? All of these questions could be answered at once if we knew the Moon's wandering within the diagram of Figure 7.8. Unfortunately this is not possible, because the effect of the tides is very difficult to apply to the long-term evolution of our satellite. Indeed, the modelling of the tidally interacting Earth–Moon system depends strongly upon the unknown values of some key parameters tied to the internal structure of the Earth.

An unexpected glimpse of the past orbit of the Moon has come from an apparently distant field of study: the analysis of the layered sediments of the Elatina formation – a Precambrian periglacial lake deposit in South Australia. It is a peculiar embayment characterised by an input of oceanic water on one side and a river inlet on the other side. The periodic variation of the level of the water due to tides at the river/lagoon interface can be recognised by the regular distribution of the grain deposition. At high tide the ocean opposes the river's flux, encouraging the deposition of light grains which remain more easily suspended and are carried onwards. Low tides exhibit the opposite behaviour, while very fine dark laminations are also present. The result is the formation of banded sediments with a complex structure and varying thickness that allows us to deduce, in detail, the behaviour of tides and therefore to compute some basic orbital parameters of the Earth-Moon system.

The reason why this geological process is so interesting is that the sediments of the Elatina formation date back to the late Precambrian era, thus representing a 'snapshot' of the lunar orbit some 680 million years ago. From their analysis – carried out by Charles Sonett and his collaborators of the University of Arizona – it has been established that the distance of the Moon was 4% less than today, and that the length of the terrestrial day was only about 22 hours. Relying on these data it has been possible to estimate that the Precambrian Moon was satisfying a Saros-like commensurability ($512\ T_S = 547\ T_A = 552\ T_N$), although the cycle had a much longer duration with respect to the present cycle: 39 years 3 days 12 hours. Another 'lucky chance'?

MOONSHADOWS

Many planetary systems have been discovered around other stars, thus contributing to the longstanding problem of understanding whether life in the Universe is an almost inevitable process or the result of a sequence of extremely low-probability events (the 'lucky chances' mentioned above). The puzzling interrelations between the peculiar orbital characteristics of the Moon and our own existence call for tentative explanation.

The large mass of the Moon, when compared to that of the Earth, must first be explained, and this anomaly must be traced back to the early times of planetary formation. Three basic hypotheses about the origin of our satellite are sensible: *cosmogonic* – Earth and Moon accreted together as independent bodies, like the majority of natural satellites; *capture* – the Moon formed somewhere else in the Solar System and was only later trapped in the gravitational field of our planet; and *collisional* – a catastrophic impact with a large wandering object caused the break-up of the proto-Earth. The latter hypothesis seems to have gained credence by now, and it is therefore used for starting a brief history of the Moon.

The year 4,500,000,000 BC is running, only some hundred million years after the solar nebula started to condense matter into solid bodies. The Earth is a hot, almost fluid body in which the heavy elements sink toward the nucleus and the rocky crust begins to cool down. A wandering protoplanet – possibly the size of Mars – happens to follow a collision trajectory with the Earth. The cosmic impact has dramatic consequences: the Earth splits into two major components. The smaller one, after attaining a spherical shape and rapidly cooling, becomes our Moon.

A few hundred million years later, a dense, acid atmosphere, due to heavy outgassing from the interior, surrounds our planet. On a smaller scale the same happens to the Moon, but our satellite does not have sufficient gravity to retain the gaseous molecules, and it soon becomes an airless world. The extended lunar seas and the densely cratered highlands reveal that the only geological activity in the Moon's history was induced by large impacts, releasing enough energy for large-scale melting of the surface material, or by steady cratering events.

At about this time, amino acids – the first complex organic compounds – began to appear on our planet. They are the result of electrostatic discharges in the atmosphere – a type of lightning – or, according to a more exotic hypothesis called *panspermia*, they come from outer space, frozen inside impacting comets. Whatever the origin, amino acids are still far from being life. The next step is to build proteins. And here comes the Moon. One of the proposed mechanisms relies on the effect of tides, which by periodically changing the sea level in shallow waters could provide the required alternative protection against exposure to the hard ultraviolet radiation from the Sun. The frequency of chemical 'mutations' is accelerated without the risk that the new compounds will to soon be destroyed by the agent that generated them.

We now move forward in time to about 3 billion years ago, when primordial life-forms already populate the Earth. Evolution through natural selection has just begun, but it will be a long time before the evolution of complex beings: plants, animals, and eventually humans. Only time? In discussing spin–orbit resonances we emphasised (in Chapter 4) the importance of the Moon in stabilising the obliquity of the Earth – the inclination of the rotation axis of our planet, which rules the seasons' cycle. Again, the Moon could have provided the long-term climatic stability needed to take evolution into the fast lane.

Since ancient times the Moon has been worshipped as a celestial goddess governing the life cycle on our planet. Birth, death and fertility are associated with the motions of our satellite, as witnessed by the widespread traditional beliefs linking the lunar phases to crop harvesting or to the frequency of pregnancies, as well as many others. By gathering astronomical information, celestial mechanics and life science it has been possible to establish a tentative scenario of the origin and the development of life on Earth, where the Moon plays a central role. Science is slowly unveiling the true motivations for being thankful to the Moon.

A lunar comedy

One of the major problems concerning Earth–Moon tidal evolution is that if we extrapolate the distance of the Moon backwards at the rate at which our satellite presently drifts away from our planet, it collides with the Earth much too soon to be consistent with any reasonable formation scenario. Relying on this fact, the celebrated Italian writer Italo Calvino, in his collection of short novels *Cosmicomics* – dedicated to providing imaginary solutions to real scientific riddles – has depicted the following fantastical, surreal and amusing scenario on 'The Distance of the Moon' (1968):

> 'How well I know! – old Qfwfq cried – the rest of you can't remember, but I can. We had her on top of us all the time, that enormous Moon: when she was full – nights as bright as day, but with a butter-coloured light – it looked as if she were going to crush us; when she was new, she rolled around the sky like a black umbrella blown by the wind; and when she was waxing, she came forward with her horns so low she seemed about to stick into the peak of a promontory and get caught there. But the whole business of the Moon's phases worked in a different way then: because the distances from the Sun were different, and the orbits and the angle of something or other, I forget what; as for eclipses, with Earth and Moon stuck together the way they were, why, we had eclipses every minute: naturally those two big monsters managed to put each other in the shade constantly, first one, then the other... Orbit? Oh, elliptical, of course: for a while it would huddle against us and then it would take flight for a while. The tides, when the Moon swung closer, rose so high nobody could hold them back. There were nights when the Moon was full and very, very low, and the tide was so high that the Moon missed a ducking in the sea by a hair's breadth; well, let's say a few yards, anyway. Climb up on the Moon? Of course we did. All you had to do was row out to it in a boat and, when you were underneath, prop a ladder against her and scramble up.'

HIGHWAYS TO THE MOON

A long-lasting heritage of the Apollo missions is represented by the mirror reflectors left in various locations on the lunar surface, for performing lunar laser ranging. This consists of sending a laser beam from the Earth's surface, aimed at one of the mirrors and detecting the return of the light after reflection. From the time taken by a radiation pulse to complete the Earth–Moon–Earth round-trip it is possible to measure the distance of the Moon to an accuracy of within a few centimetres.

The last humans to walk on the Moon left there in December 1972, after the Apollo programme had successfully completed six missions on the surface. The astronauts left on our satellite many of the scientific instruments that they had used, several pieces of space technology such as the lunar rovers that they driven on the dusty lunar surface, and brought back hundreds of kilogrammes of lunar rock samples. The next step for lunar exploration appeared straightforward: the establishment of a lunar base where humans could live for extended periods of time – the first settlement on an alien world. The vision was strongly supported by the leader of the Apollo project, Wernher Von Braun (1912–1977), who declared it feasible before the end of the 1970s. But his hopes were frustrated by a sudden change in US strategy, which focused on the development of the Space Shuttle, providing routine access to low Earth orbit for commercial and military purposes. The Moon was even removed from the list of celestial bodies targeted by unmanned spacecraft. The situation remained unchanged for more than a decade, until in 1990 a tiny Japanese probe – named Hiten, after a music-playing Buddhist angel – flew past the Moon several times.

The importance of Hiten was not related to science (it carried almost no instruments) nor in having finally marked a return to lunar exploration. The breakthrough was for celestial mechanics. However, the orbital path of the spacecraft was completely different from a classical Earth–Moon transfer trajectory (an ellipse with apogee sufficiently distant that the spacecraft is affected by the gravitational pull of our satellite) (Figure 7.9). Hiten's innovative trajectory derived from advances in the dynamical systems approach to the three- and four-body problems (discussed in Chapter 2). When applied to the Earth–Moon–Sun–spacecraft system, new ways to the Moon were been found.

In particular, the chaotic regions associated with the collinear Lagrangian point L_1 are exploited to 'gravitationally' redirect the motion of the spacecraft, instead of firing the onboard propulsion system. The resulting trajectories passing through the 'fuzzy boundary' close to the Lagrangian points are called *weak stability boundary* (WSB), since they are able to ride the instabilities associated with an orbital motion close to the boundary between the gravitational domains of the Earth and the Moon or of the Earth and the Sun (Figure 7.10).

At first glance the WSB trajectory displayed in Figure 7.11 appears to be counter-intuitive: a long tour reaching up 1.5 million km in the direction of the Sun is performed before going back to the much less distant lunar orbit (380,000 km).

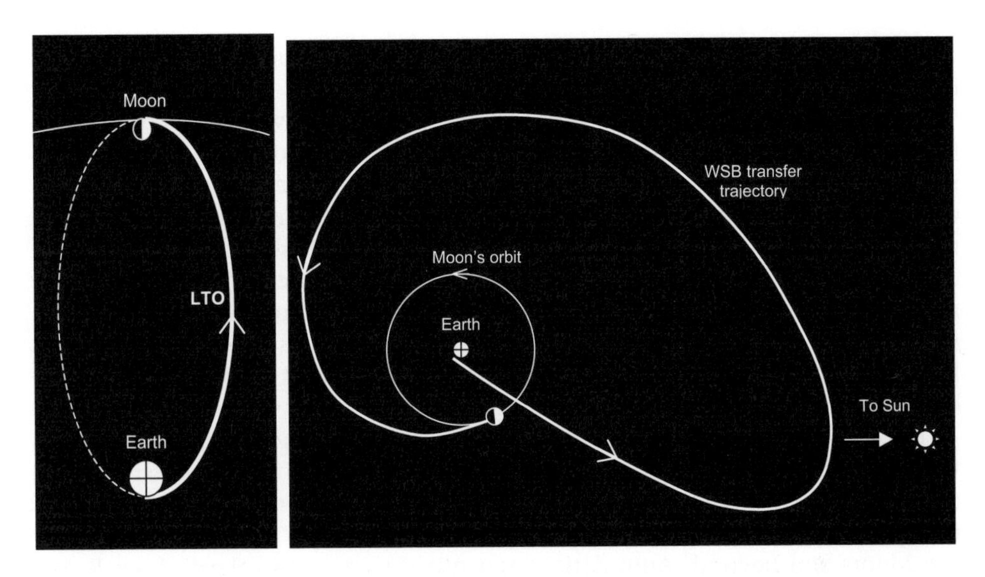

FIGURE 7.9. Traditional transfers to the Moon (left) obtained by larger eccentricity ellipses, referred to as lunar transfer orbit (LTO). WSB transfers exhibit a far more complex orbital path, as shown for the Japanese Hiten spacecraft (right).

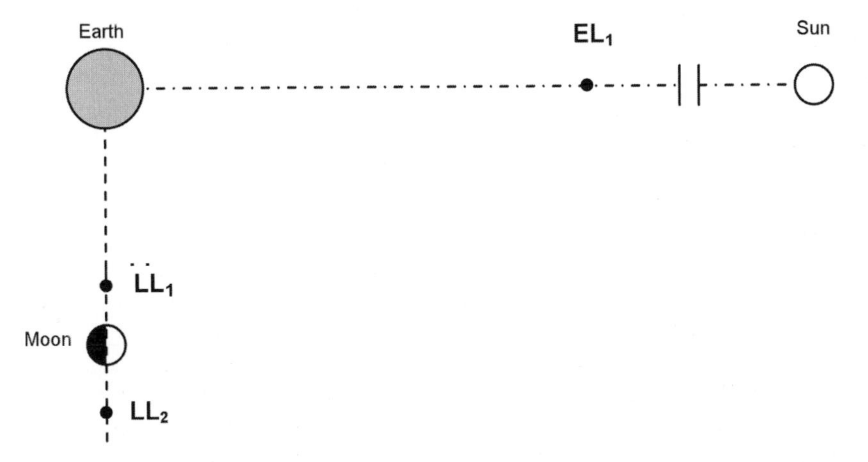

FIGURE 7.10. WSB trajectories exploit the chaotic regions associated with the collinear Lagrangian points. Three of them are of particular significance for lunar exploration. The closest to our planet lies along the Earth–Moon direction 60,000 km from our satellite, and is usually indicated by LL_1 (lunar Lagrangian point L_1) in order to distinguish it from the Lagrangian point L_1 of the Earth–Sun system (EL_1), which is located about 1.5 million km away in the direction of the Sun. Both of them have been used for space mission design – the former for taking the European SMART-1 spacecraft to the Moon, and the latter by the Hiten mission. The translunar libration point LL_2 could be used either for a halo orbiter (see Chapter 2) or as a gateway for accessing interplanetary space.

The transfer time increases accordingly. The Apollo missions reached the Moon in a few days, while Hiten travelled for 100 days or so.

Why, then, should this trajectory be used? The advantage is that WSB trajectories exploit the perturbations of both the Moon and the Sun to achieve a selenocentric orbit, thus sparing some of the fuel otherwise needed for the lunar orbital insertion manoeuvre. As less propellant is required, more of the spacecraft mass can be devoted to the payload. This is a remarkable result for science, as it allows an increase in the number of scientific instruments flown, thus also improving the scientific return of the mission. It is even more important if the vision of a permanent human presence on the Moon is resumed, because of the large amount of resources (food, fuel, infrastructure, and so on) that must be continuously sent from the home planet.

The project of establishing an outpost on the Moon is now firmly back in the front line, representing the core of the American vision for space exploration (Figure 7.11). The baseline scenario foresees the development of a new space transportation system, the Crew Exploration Vehicle (CEV), capable of reaching the Moon and beyond. Automatic cargo missions are also planned for heavy loads, exploiting electric propulsion engines, which are much more efficient than chemical propulsion, although their low thrust results in longer transfer times.

A lunar base hosting twelve astronauts could be ready before 2020. The proposed name is Jamestown-on-the-Moon, after the first English settlement in North America. It will presumably be located at the south pole, since those regions fulfil two basic requirements: an extended period of illumination during the lunar month, and the proximity to craters with regions permanently in the shadow, as these have probably preserved large amounts of water ice. In the context of the Moon base, a number of technological and scientific infra-

FIGURE 7.11. An historical footprint left by an Apollo astronaut on the Moon, and an artist's view of a future human settlement on the Moon. (Courtesy NASA.)

The hammer and the feather

A striking example of the advantages of carrying out scientific experiments on the Moon was provided by a simple experiment performed during the Apollo 15 mission (Figure 7.12). According to gravitation theory, falling bodies are subjected to the same acceleration, whatever their mass or dimensions. A hammer and a feather must reach the ground together when departing from the same height. It is not easy to check this result on Earth due to the presence of the air, which slows the fall of the lighter object; but the Moon is perfect for a live demonstration. The following experiment was carried out by Commander David R. Scott on 3 August 1971, and was recorded by the cameras:

> 'Well, in my left hand I have a feather; in my right hand, a hammer. And I guess one of the reasons we got here today was because of a gentleman named Galileo, a long time ago, who made a rather significant discovery about falling objects in gravity fields. And we thought where would be a better place to confirm his findings than on the Moon? And so we thought we'd try it here for you. The feather happens to be, appropriately, a falcon feather for our Falcon [the lunar descent module]. And I'll drop the two of them here and, hopefully, they'll hit the ground at the same time. [The hammer and feather fall side by side and touch the ground simultaneously.] 'How about that! Which proves that Mr. Galileo was correct in his findings.'

FIGURE 7.12. Astronaut David R. Scott, Commander of Apollo 15, holds a feather in his left hand and a geological hammer in his right as he prepares to drop them in a test of Galileo's law of motion concerning falling bodies. The two hit the lunar surface simultaneously. (Courtesy NASA.)

structures are also envisaged. The operation of large solar power plants for producing energy to be sent to Earth in the form of electromagnetic radiation could provide the long-sought energy independence of our planet. A condominium of astronomical observatories on the Moon could explore the Universe at all wavelengths (from X-rays and gamma-rays to the optical, infrared, microwave and radio regions) without the limitations due to atmospheric disturbances. Exploitation of *in situ* resources will probably lead to the production of oxygen-based propellants to be used for lunar-based long-range exploration of the Solar System.

All these activities rely on various levels of advanced spaceflight dynamics. Short-duration Earth–Moon trajectories are mostly required for manned missions because of the need to protect the crew from exposure to energetic radiation such as that produced by sudden and unpredictable solar flares. Long-duration LL_1 (Lunar Lagrangian point L_1) and EL_1 (Earth Lagrangian point L_1) WSB transfers, using electric propulsion, maximise the efficiency of unmanned cargo systems, thus fulfilling at best the routine maintenance of a lunar base. Safety equipment, refurbishment and spare parts stationed in an LL_1 halo parking orbit would, on demand, reach any point on the surface of the Moon within a few hours. A data-relay satellite placed on the translunar LL_2 point would provide global coverage of the far side of the Moon in order to avoid communications breakdown of eclipsing circumlunar spacecraft, and to ensure continuous radio links with the Earth for scientific infrastructures such as radio telescopes.

The chances that all of this will be realised seems to be rather high. Since the 1960s the worldwide political situation has evolved, and there is no need of a technological space race for affirming supremacy. Moreover, the number of nations with independent access to space has grown considerably: Europe, Japan, China and India have current or under-study advanced lunar exploration plans. The colonisation of the Moon will most probably be carried out as a joint effort by many different countries, sharing the risks and the benefits of the challenge. Hopefully, we will be soon return to the Moon.

8

Rock around the planets

> One of our problems is trying to figure out which way is up and which way is down.
>
> *John Young, Apollo 10*

From a strictly dynamical point of view the difference between a natural and an artificial celestial body relies on the possibility that the latter has to follow trajectories chosen not only by gravitation but also according to human will. *Spaceflight dynamics* is a branch of celestial mechanics that has developed along with the advances in the astronautical sciences for controlling the motion of artificial satellites and interplanetary spacecraft. Orbital changes are obtained either by properly operating an onboard propulsion system or by exploiting gravitational perturbations, thus opening new and exciting perspectives for celestial mechanics. In particular, the direct exploration of all the major Solar System bodies has been made possible by deep-space probes following complex trajectories, where manoeuvres and close encounters with the planets play a deciding role in sending a spacecraft toward its target. This is why spaceflight dynamics should not be considered only a practical application of known principles, but also as a novel approach to orbital motion.

SPACE IN FLIGHT

The history of flight dynamics raises mixed emotions: the pride and glory of paving the way for humans landing on the Moon and returning safely to Earth; and the shame and fear of sharing the same technology which led to the development of weapons of mass destruction, from the German V2 rockets to modern nuclear missiles. Unfortunately there is no way of separating the scientific, commercial or military nature of orbital motion, as the aim of spaceflight dynamics is to find the trajectories which best satisfy a given mission profile.

In this respect the possibility that a spacecraft has to modify its orbital path by firing an onboard propulsion system can be considered as a natural extension of the concept of perturbation. It is a different problem with respect to classical celestial mechanics – not just in modelling perturbations in order to account for the observed motion of the celestial bodies, but also in computing the perturbations needed to send a spacecraft on a desired orbital path. Manoeuvring an artificial satellite or a deep-space probe is a difficult business. A spacecraft is mainly influenced by the gravitational pull of the celestial bodies, the magnitude of the gravity being dependent on their mass and distance. However, its motion also depends on non-gravitational forces such as solar radiation pressure and atmospheric drag, as well as on the characteristics of the onboard rocket engine (high-thrust chemical or low-thrust electric) used for performing orbital manoeuvres.

In particular, the efficiency of space-qualified propulsion systems is far from being comparable to their analogues used for transportation on our planet. Half of the mass of a spaceship travelling to Mars is the fuel needed upon arrival for 'parking' around the planet (the orbit injection manoeuvre). If this were so for the engine powering an ordinary car, a dedicated trailer would be needed to host a fuel tank large enough to provide a sufficiently long range. The additional bad news is that there are no fuel stations in space, and the propellant needed for all the orbital manoeuvres planned for the entire lifetime of a mission must be loaded onto the spacecraft before departure from Earth. The mass budget of a spacecraft is therefore a true nightmare for mission analysts, faced with the difficult task of reconciling 'heavy' commercial or scientific goals with the laws of celestial mechanics. Trajectory optimisation programmes have been developed to this end. Highly sophisticated computer modelling of the gravitational environment and of the execution of manoeuvres allows us to search, among the many orbital paths fulfilling the mission requirements, for those ensuring major fuel savings.

Time also runs at a different pace for spaceflight dynamics. Before the space age, celestial mechanics was mostly concerned with orbital evolution covering astronomical timescales. As discussed in the previous chapters, the stability of the Solar System is measured in billions of years, and even the chaotic transport of matter from the asteroid main belt needs millions of years to be effective. Due to the limited energy budget available and to the onboard components' consumption, a spacecraft has a much shorter operational lifetime of the order of ten years. As a consequence, orbital configurations that in the long term are unstable, might happen to be a convenient choice when dealing with the short lifetimes of space missions. In this context, some peculiar orbital regimes have been discovered, during mission analyses, only recently, since there were no celestial bodies that could dynamically survive in those configurations for a sufficiently long timespan to draw the attention of the astronomers.

This is the case of the *halo orbits* around the collinear Lagrangian points, which were first found during flight dynamics studies carried out for the Apollo programme. Indeed, one of the most critical events of the whole mission was the communication black-out when the Apollo capsule flew behind the Moon. During this period the astronauts were completely alone, almost 400,000 km from home, without any means of communicating with the mission control centre in case of emergency. The only possibility for overcoming the problem

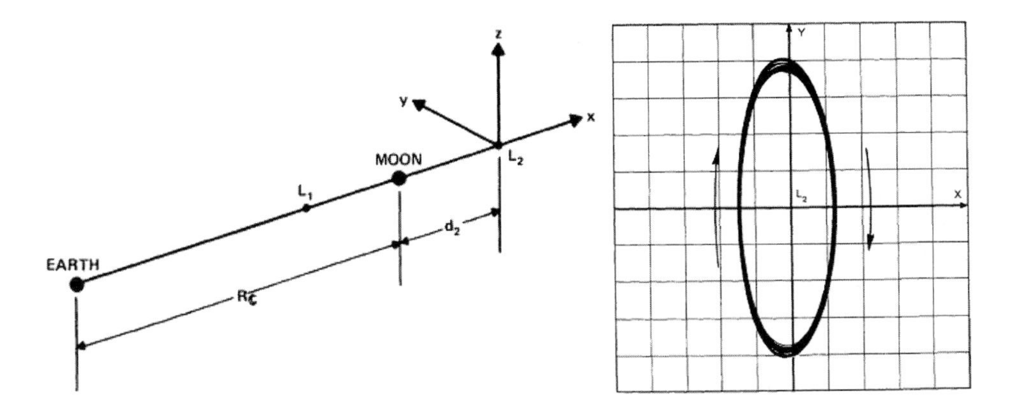

FIGURE 8.1. As seen from the Earth, in an Earth-centred reference frame rotating with the same angular velocity of the Moon (left plot), a halo orbit around the translunar point traces an elliptical path in the sky (right plot). The width of the halo can be tuned in order to be always larger than the apparent diameter of the Moon, thus allowing a data-relay satellite to be always in view of both the Earth and the far side of the Moon. The plots shown are those originally included in the article by Farquhar and Kamel, who first pointed out their importance.

was the availability of a data-relay telecommunication satellite satisfying the double feature of being always in view from both the Earth and from the far side of the Moon; but an orbital configuration satisfying these peculiar constraints was not mentioned in the vast literature of celestial mechanics. However, in 1973 a seminal paper, coauthored by an outstanding member of the Apollo team, Robert Farquhar, showed the existence of quasi-periodic orbits characterised by large-amplitude librations around the translunar equilibrium point L_2 (recall that the collinear Lagrangian points are three equilibrium positions lying in the Earth–Moon direction. One of them – denoted by L_1 – is between the Earth and the Moon, while the other two positions – denoted by L_2 – are respectively behind the Earth or behind the Moon.) As seen from Earth, the librations around L_2 appeared as elliptically-shaped paths around the Moon; thus the term 'halo orbits' (Figure 8.1).

Although an L_2 halo orbiter could definitely avoid the telecommunication breakdown suffered during the Apollo missions, the project was never realised due to financial reasons. Yet the use of halo orbits is now well established. Solar missions such as SOHO are particularly suited for orbiting around the Earth–Sun L_1. The advantage is two-fold: a continuous monitoring of solar activity on one side, while 'haloing' the Sun to avoiding transiting the solar dish which produces electromagnetic disturbances that inhibit the radio link with Earth. On the other hand, the collinear Lagrangian point behind our planet along the Sun–Earth direction is a perfect location for space telescopes. By always pointing in the antisolar direction, the sky can be observed without the worry that the Sun, the Earth or the Moon will appear in the field of view of the telescope, while the whole celestial sphere is covered in the course of one year, as the Earth revolves around the Sun.

ORBITING THE EARTH

There is a little celestial mechanics in our everyday life. A stone tossed in the air is subjected to the same force that drives the motion of planets, stars and galaxies, and the resulting trajectory is therefore a branch of ellipse or hyperbola – an orbital path broken by the intersection with the surface of our planet (Figure 8.2). It is therefore not surprising that almost three centuries before the first artificial satellite was launched, Isaac Newton could already imagine how to send a stone in orbit around the Earth: 'The greater the velocity with which it is projected, the further it goes before it falls to the Earth... till at last, exceeding the limits of the Earth it will pass into space.'

The reason why it was not possible for Newton to throw a stone into orbit is that the corresponding velocity is extremely high – around 30,000 km/h – and it is obvious that no-one could do it before the invention of rocket propulsion. This value is computed by applying some basic celestial mechanics. Consider the two-body problem Earth-stone and derive the parameters associated with a hypothetical geocentric circular orbit having a radius exactly matching that of our planet (6,378 km). It is, of course, a purely theoretical exercise, but it provides the basic framework for studying the dynamics of Earth-orbiting satellites. The stone represents a *grazing satellite* – the lowest-altitude artificial satellite that could possibly be imagined. Its period of revolution around the Earth – about 1.5 hours – can be computed with elementary equations.

There are three main non-military reasons for sending an artificial satellite into orbit around the Earth: exploiting space communications, remote sensing of our planet, and carrying out science in space. Each of these goals is characterised by different mission requirements, and spaceflight dynamics has extensively applied the methods of celestial mechanics for finding the corresponding best orbital configurations.

In Chapter 4 the peculiar spin–orbit resonance characterising the geostationary satellites and the advantages for telecommunications were introduced. The possibility of keeping a satellite almost fixed in the sky as seen from the ground has allowed the development of direct TV broadcasting, as witnessed by the widespread diffusion of parabolic antennae on private houses and apartments. If the satellite were not on a circular 24-hour orbit, the ground antenna could not have an easy-to-install fixed orientation, but it should be capable of tracking the satellite's apparent motion in the sky. Moreover, signal transmission would be periodically interrupted when the satellite moves out of visible range.

Global coverage ensuring a radio link between any two points on the surface of the Earth is guaranteed only by placing in orbit constellations of satellites – identical spacecraft able to communicate with each other. Peculiar orbital

Rocket man

In the original formulation in the *Principia Mathematica* for explaining how to attain an orbital motion around the Earth, Newton used the example of firing a cannon from the top of a mountain (Figure 8.2). But even if such an extremely powerful cannon could actually be built, able to provide the required velocity for orbit insertion, a stone could never pass into space. The air resistance would immediately slow its motion, while the heat produced by friction would probably completely vapourise it. Moreover, this scenario is not applicable to manned missions, since the sudden acceleration at departure would immediately kill the astronauts. Realising Newton's dream needed a different thinking.

Konstantin Eduardovich Tsiolkovsky (1857–1935) (Figure 8.3) was born in the village of Izhevskoe, south of Moscow, and when he was 10 years old he became almost completely deaf as a consequence of a severe illness. Notwithstanding this handicap, he studied mathematics and physics, mostly on his own. Inspired by the growing expectations from scientific discoveries (Mendeleev had just devised his periodic table of the elements) he developed the first design for a multi-stage rocket, which later became the standard means to launch artificial satellites. Tsiolkovsky understood that the velocity required to achieve orbital motion should be provided during the flight, and not simply at the beginning. He also showed that the action–reaction principle could be exploited. When an object is thrown in a certain direction, a push is applied in the opposite direction (for example, the hard pressure on the shoulder of someone firing a gun).

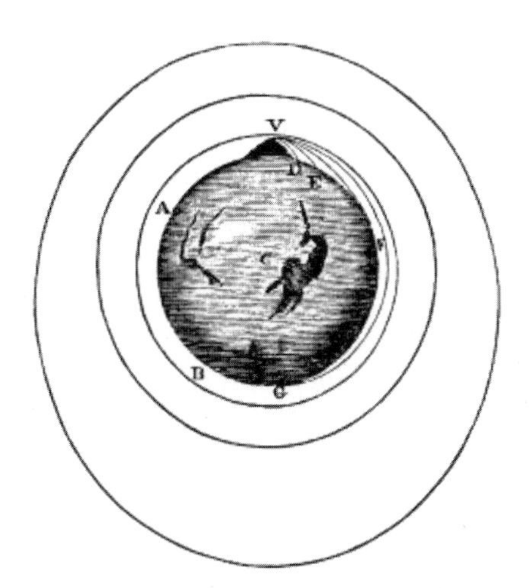

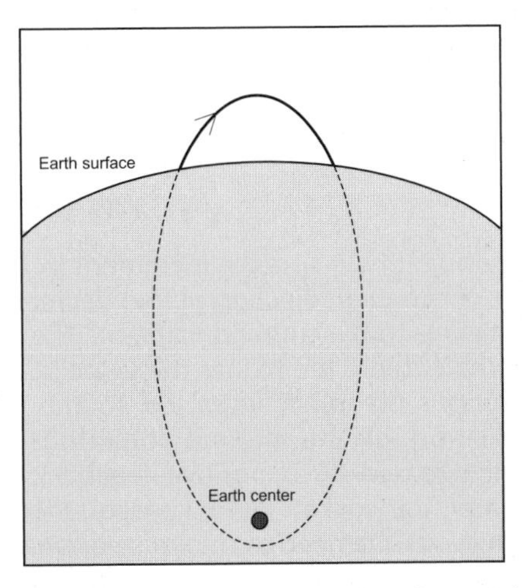

FIGURE 8.2. Isaac Newton's original drawing tracing back to the same physical principle the trajectory of a cannon ball fired from the top of a mountain, and orbital motion (left). The diagram at right shows the branch of ellipse (centre) described by a stone thrown from the surface of the Earth.

FIGURE 8.3. Konstantin Tsiolkovsky, and the spectacular launch of the Space Shuttle carrying the Galileo spacecraft bound for Jupiter. (Courtesy NASA/JPL.)

Chemical rocket propulsion is achieved by the explosive burning of fuel, producing exhaust gas and particles that are expelled at high velocity by the rear nozzle of the engine, thus pushing the rocket in the opposite direction. In some sense, Newton's cannon has to be on board and not on the ground – and must be continuously fired. The steady acceleration provided by the principle of action–reaction increases the velocity of the rocket until a desired orbit is achieved. In order to compute the efficiency of this process, Tsiolkovsky produced his celebrated rocket equation:

$$V - V_0 = v_e \ln (m_0/m).$$

The quantity $V - V_0$ is the increment in velocity obtained when the burning and the consequent ejection of fuel decreases the total mass of the rocket from the initial m_0 to the running value m. The exhaust velocity v_e measures the power of the engine: the higher the velocity with which the fuel mass is ejected, the stronger the push imparted to the rocket. The appearance in the rocket equation of the natural logarithm (ln) indicates that the dynamics is characterised by a positive feedback: the larger the fuel mass burned, the lighter the rocket. As a consequence, a higher velocity increment can be achieved. The need to minimise onboard mass has led to the concept of multi-stage rockets, in which the fuel tanks are jettisoned during the flight, contributing to a further reduction in mass.

configurations are then studied in order to ensure that at least one satellite is always in view from both the ground and from another satellite. Constellations are also used for navigation purposes, such as the well-known Global Positioning System (GPS) and the European Galileo project aimed at building a global navigation satellite system. They are characterised by intermediate-altitude orbits (about 25,000 km), and will soon allow ground positioning with an accuracy of less than 1 metre.

Monitoring the health of our planet from space is becoming an increasingly important issue for security and civil protection. The thinning ozone layer over the poles has been observed from space, and is now kept under control by means of dedicated Earth-observation missions. Meteorological satellites - traditionally used for weather forecasting - represent one of the first space applications entering everyday life. Modern remote sensing satellites provide high-resolution images of the Earth's surface, and are therefore usually on high-inclination low-altitude orbits (below 1,000 km). In order to provide full coverage of the Earth's surface or the continuous monitoring of a specific region, peculiar resonant orbits are chosen to guarantee the periodic repeatability of the satellite ground tracks. Perturbations due to the non-spherical shape of the Earth are also fruitfully exploited, and we can take advantage of the slow nodal precession induced by the oblateness of the Earth so that the orbital plane of a satellite always remains perpendicular to the Earth–Sun direction. These Sun-synchronous orbits allow us to image the surface of our planet always with the same illumination conditions, thus allowing meaningful comparisons among different observations.

Scientific missions can also be found on classical low-inclination low Earth orbits (LEOs), such as the 500-km-altitude circular orbit where the Hubble Space Telescope is operated. An example of a more exotic dynamical configuration is provided by the European Cluster mission, by which the need for simultaneous probing of different regions within the Earth's magnetosphere is obtained by means of four independent spacecraft on polar elliptic orbits, kept in a tetrahedral configuration with relative distances ranging from 600 to 20,000 km. Because of this peculiar 'group dancing' around the Earth, the Cluster spacecraft have been given the names Rumba, Salsa, Samba and Tango.

SPACE DEBRIS

The first artificial satellite to orbit the Earth closely resembled Newton's cannonball. The Russian Sputnik was a 50-cm aluminium sphere with long outstretched antennae. Launched on 4 October 1957 into an elliptical orbit ranging from 215 to 940 km, it marked the beginning of the space age. Since then, the number and dimensions of artificial satellites has steadily grown, and today the launching of commercial and scientific satellites is routinely performed by American, Russian, European, Japanese, Chinese and Indian rockets. The robust nature of space technology has grown accordingly, allowing the safe functioning of a satellite over 10 years or more. Yet starting from an altitude of about 700 km the Earth's atmosphere is so thin that the friction slowing down the motion of an artificial satellite and causing eventually its re-entry can be considered as negligible. Therefore any object flying above this limit suffers almost no atmospheric drag and is therefore bound to orbit the Earth on timescales much longer than its operational lifetime.

It might be wondered, then, whether this accumulation of material might give rise to some sort of 'cosmic traffic jam' due to the uncontrolled use of space. The problem is real, and is worsened by two additional considerations. During the early phases of a mission (launch, satellite separation from the rocket, deployment, activation of the onboard instruments, and so on) a number of minor components are ejected into space, these also becoming, for celestial mechanics, artificial satellites. Launch failures also provide a substantial contribution. For example, the explosion of the third stage of a rocket or of a spacecraft fuel tank generates thousands of fragments which continue to orbit the Earth for a rather long time. All these objects are now referred to as 'space debris' (Figure 8.4) – a population of celestial bodies that must be kept under strict investigation and control, as it represents a major threat to present and future exploitation of near-Earth space. Collisions with operational satellites or manned spacecraft can have dramatic consequences because of the high relative impact velocities, with values of the order of a few kilometres per second (1 km/s = 3,600 km/h) and with an energy release typical of a hypervelocity projectile.

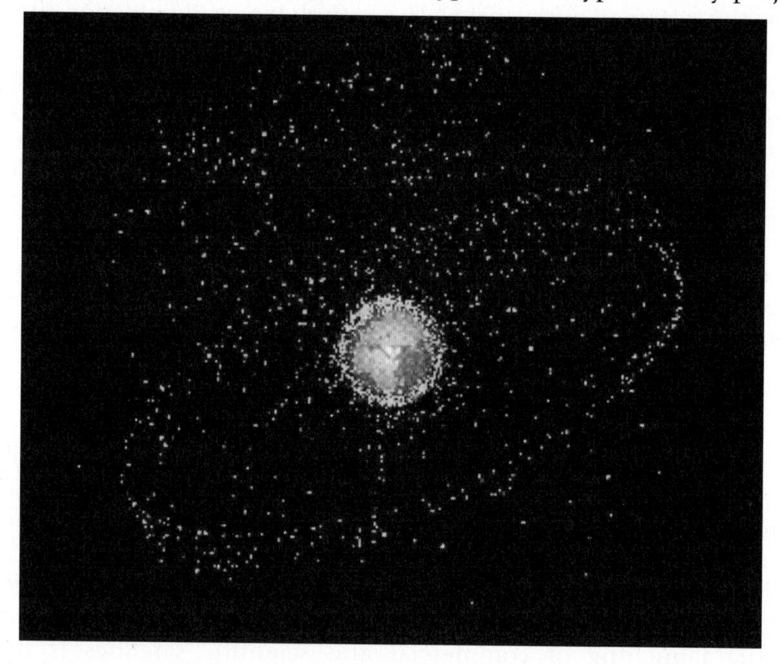

FIGURE 8.4. The position of all catalogued debris orbiting the Earth. A higher density can be recognised along the geostationary ring and where high-inclination polar orbits are located. (Courtesy ESA.)

The problem began to be taken seriously at the beginning of the 1980s, because of the increasingly crowded geostationary ring; but it was soon recognised that it was even more important for intermediate and LEO orbits, especially in view of the realisation of satellite constellations. One of the first consequences was the thickening of the protective shields built for the International Space Station modules, and the restrictions posed on the Space Shuttle's flying attitude, in order to always expose the lowest possible area to the incoming flux of space debris. Radar and optical observation campaigns were started from the ground for the discovery and follow-up of debris, while modelling of the future evolution of the whole population is actively developing.

Drops in the sky

When looking at the catalogue of space debris, an anomalous concentration of objects at an altitude between 850 and 1,000 km is detectable. Detailed investigations by radar observations identified a densely populated 'family' of debris on circular 65-degree inclined orbits, and it was discovered that there are about 60,000 such particles larger than 8 mm and a few hundred larger than 3 cm. They share not only the dynamics but also the physical properties, which are consistent with droplets of liquid sodium–potassium (NaK). But what were the circumstances that generated such an unusual and dangerous swarm?

After extensive analyses, both the USA and Russia came to the conclusion that the source was a series of nuclear-powered Russian radar ocean reconnaissance satellites (RORSAT). In order to avoid contamination in case of an uncontrolled re-entry in the atmosphere due to a malfunction (as happened in January 1978 when Cosmos 954 dispersed radioactive waste over a sparsely inhabited area of Northern Canada), a safety strategy was implemented. At the end of the operational lifetime of each RORSAT satellite, the nuclear reactor was boosted to a high altitude graveyard orbit, thus preventing orbital decay. The problem was the way in which this was done, because during the abrupt separation a leakage of the reactor coolant liquid (NaK) occurred, producing a swarm of particles. Even if not large enough to cause the catastrophic disruption of a spacecraft, the dangerous particles considerably increase the probability of small cratering events which can produce significant damage to operational satellites.

At present about 10,000 objects with a diameter larger than 20 cm are known to orbit the Earth, but only 350 are operational satellites. When accounting for millimetre-sized debris, there is a total of 3 million objects.

The collisional evolution of the space debris population has many similarities with the studies of the long-term dynamical evolution of the asteroid main belt or of Saturn's rings. They share an identical scenario in which each collision produces new potential impactors by fragmentation of the target, thus increasing the overall population density. As far as near-Earth space is concerned, the risk is a runaway growth of the debris within a few tens of years, thus increasing the collision probability to a non-tolerable level. In this respect our planet will soon be surrounded by 'artificial rings' – regions where the commercial, human and scientific exploitation of space will be inhibited for a long time.

The space debris hazard has been certainly underestimated in the past, and the urgent response of the international community is now required. There is little point in trying to 'clean' the sky, as the removal of this debris is neither technically nor economically feasible. Mitigation rules – such as the strict limitation of components jettisoned in space, deorbiting (controlled re-entry into the atmosphere), and moving a satellite, at the end of its operational lifetime, toward previously identified graveyard orbits – are being increasingly adopted. The hope is that they will be effective and be applied worldwide without exception. The worst predictions foresee that the region between 700 and 1,000 km will be saturated within 50 years.

Meanwhile, the first confirmed collision between two catalogued space objects was recorded in July 1996, when the Cerise microsatellite suddenly went out of control. Detailed investigations led to the conclusion that it was hit by a fragment of the European Ariane rocket at a relative velocity of 15 km/s. Fortunately the damage involved only a subsystem, while the main body of the satellite survived the impact – otherwise a dangerous swarm of fragments could have been generated.

THE ACCESSIBILITY OF CELESTIAL BODIES

One of the most well-known quotations from Konstantin Tsiolkovsky is: 'Earth is the cradle of humanity, but one cannot remain in the cradle forever.' By studying rocket propulsion Tsiolkovsky took the first steps for enabling humans to climb up the gravity well of our planet and go into space. But leaving the cradle implies being able to travel the spaceways to other worlds, at distances farther away than the Moon, and on trajectories subject to completely different gravitational environments.

This problem was enthusiastically approached by Walter Hohmann (1880–1945) at the beginning of the twentieth century. Hohmann tried to establish the implications of planning a voyage to Venus or Mars. His reference scenario was that of a manned mission, and he soon realised that an overwhelming amount of fuel would be required for manoeuvring a spacecraft in interplanetary space. This led him to conclude that a proper choice of the trajectory is essential for the feasibility of missions into space. To this end he introduced the concept of the 'accessibility' of a celestial body as measured by the so-called ΔV – a quantity that provides the change of velocity required to reach a desired target.

Using simple two-body approximations, Hohmann showed that among all possible orbital paths joining the Earth with the planets there are particularly favourable transfer trajectories which allow for a considerable decrease in ΔV. Many decades later it was demonstrated that in most cases these trajectories are also optimal trajectories, in the sense that no other orbital path is characterised by a lower ΔV.

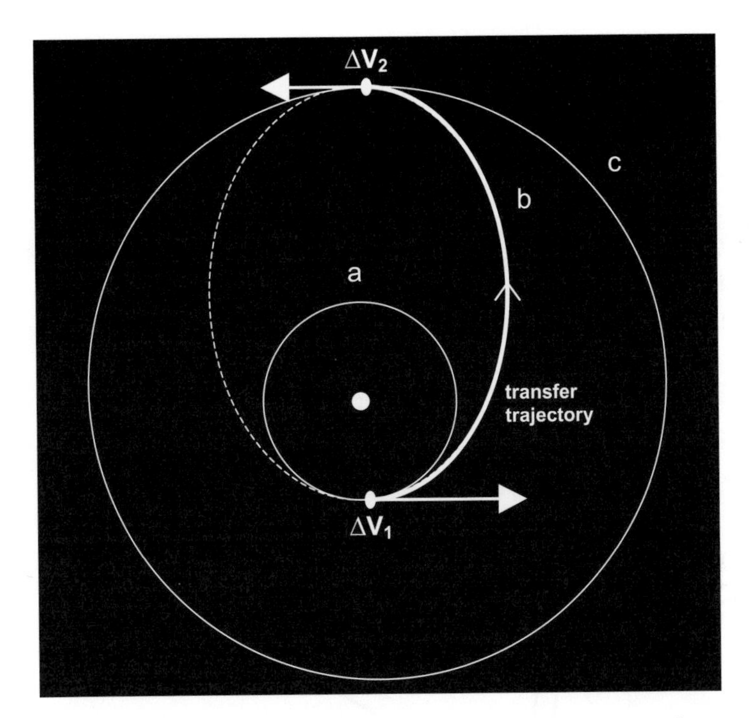

FIGURE 8.5. Orbit diagram of a Hohmann transfer. The spacecraft, initially on a circular orbit (a), is injected by an impulse of magnitude ΔV_1 into a transfer ellipse (b) with apocentre equal to the radius of the target orbit (c). Upon reaching it, a ΔV_2 of the exact amount needed for circularisation is performed. The accessibility of a celestial body is then given by the sum $\Delta V = \Delta V_1 + \Delta V_2$.

Consider the general case of a transfer between two circular, coplanar orbits, and assume that the radius of the target orbit is larger than that of the departure orbit. Suppose that a celestial body is located on the target orbit. The Hohmann strategy foresees two orbital manoeuvres, which must be performed by firing an onboard propulsion system able to provide the required impulse. The first manoeuvre injects the spacecraft into a transfer trajectory with the apocentre tangential to the target orbit (Figure 8.5). If no other action is taken and the timing is properly chosen, this corresponds to a fly-by mission profile. After a short close encounter with the celestial body on the target orbit, the spacecraft drifts away, following the descending branch of the Hohmann transfer ellipse.

The second Hohmann manoeuvre is applied upon reaching the apocentre of the transfer orbit and has the goal of raising the pericentre of the transfer ellipse until a circular orbit is again achieved. When this is done the spacecraft has the same velocity as the target body, a basic requirement for leading the spacecraft to orbit around the target body, thus accomplishing a rendezvous mission.

Adding up the ΔV contribution of both manoeuvres, a reliable estimate of the energy needed for performing a rendezvous mission is obtained. The reason

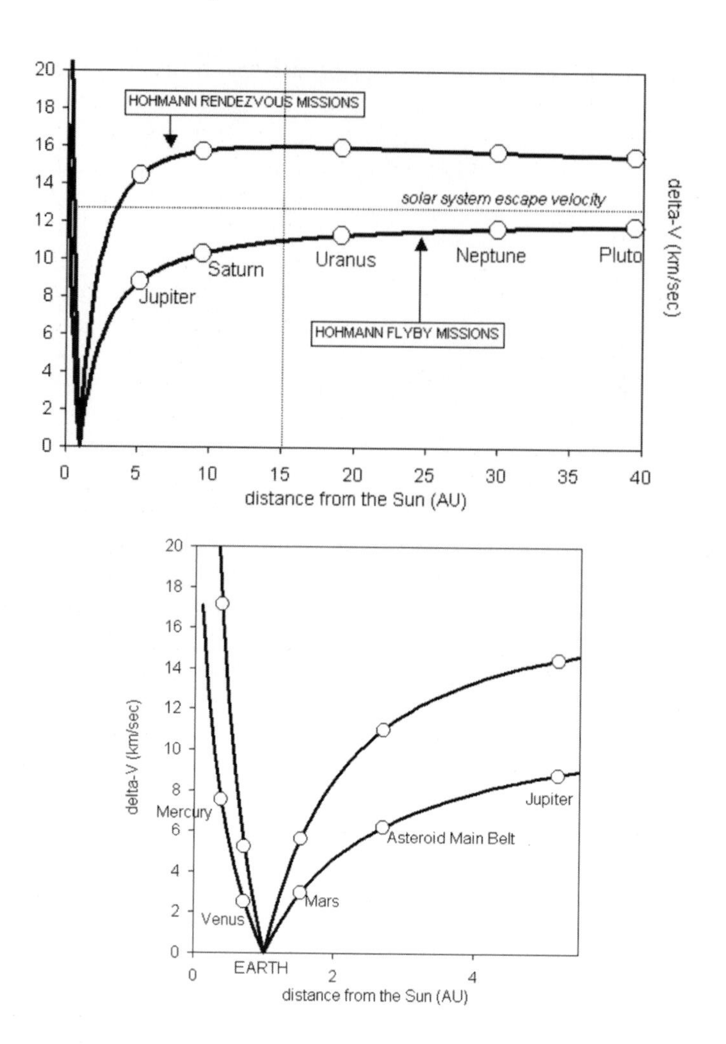

FIGURE 8.6. (top) An H-plot is a graphical representation of the Hohmann accessibility throughout the Solar System. The lower curve corresponds to the ΔV_1 needed to leave the orbit of the Earth on transfer ellipses of increasing aphelion (right branch) or decreasing perihelion (left branch) distances. The upper curve represents the total ΔV budget for performing rendezvous missions. The locations of the planets are marked on the graph as open circles, shown on both curves for the same value of the semimajor axis. The two curves tend to the Solar System escape velocity (12.34 km/sec). (bottom) When focusing on distances smaller than the distance to Jupiter, the extremely high requirements for reaching Mercury appear in the H-plot, while Venus and Mars exhibit similar accessibility.

why Hohmann chose rendezvous missions as representative of the accessibility of celestial bodies is that, because of the low relative velocities involved, they are the only means for ensuring a safe landing of humans on another planet.

A graphical representation of performing 'planetary' transfers (between circular coplanar orbits) throughout the Solar System is obtained using an *H-plot* (Figure 8.6). If the departure orbit is that of the Earth, by continuously varying the radius of the target orbit the Hohmann strategy translates into a diagram in which the behaviour of the accessibility parameter (the ΔV magnitude) is displayed as a function of the target distance. In each plot the lower curves display the ΔV_1 magnitude, while the upper curves have been obtained for the total ΔV and are therefore representative of the overall accessibility of the Solar System.

The H-plot is useful for visualising some of the problems that are likely to be encountered when planning a voyage to the planets. The enlargement of the V-shaped region in the bottom part of Figure 8.6 shows that the ΔV quickly reaches very high values when travelling toward the Sun. The orbit of Mercury (semimajor axis a = 0.35 AU) is relatively close to that of the Earth (a = 1 AU), but the corresponding ΔV is higher than that needed to reach Saturn (a = 10 AU). This can be explained by the heavy penalty of manoeuvring too close to the strong gravitational pull of the Sun, and it justifies the need for advanced propulsion systems.

In Table 8.1 the accessibility of the planets is summarised, showing that it does not correlate with the distance of the planets. The minimum distance is computed as the difference between the semimajor axis of the orbits, which also corresponds to the minimum achievable distance of each planet from the Earth (at conjunction). Besides the case of Mercury, it also appears that Neptune (15.7 km/s) is more accessible than Uranus (15.8 km/s). This apparent paradox can again be explained in terms of gravitation, because as we recede from the Sun its gravitational field becomes increasingly weak, and the magnitude of the circularisation manoeuvre ΔV_2 decreases accordingly. In general it can be stated that the accessibility of a celestial body results from the interplay of geometrical and dynamical factors.

Table 8.1. Summary of the basic requirements for reaching the planets from Earth.

	Minimum distance (AU)	ΔV (km/s)	Transfer time (days)	Waiting time (days)	Round trip (years)
Mercury	0.6	18.1	106	67	0.8
Venus	0.3	5.2	146	467	2.1
Mars	0.5	5.5	259	454	2.7
Jupiter	4.2	14.3	998	215	6.1
Saturn	8.5	15.5	2,209	342	13.0
Uranus	18.1	15.8	5,859	340	33.0
Neptune	29.0	15.7	11,184	280	62.0

Table 8.1 also includes another parameter which is equally relevant for space exploration: the time needed for accomplishing a given mission profile. More precisely, it shows the duration of the transfer time needed to reach a desired

Space architecture

Walter Hohmann was a civil engineer, and city architect of Essen, Germany. He became interested in space science later in his life, but with an ever-growing enthusiasm. His son Rudolf recalls that his father's passionate dedication to space travel pervaded family life, and he took any opportunity to share his vision of a future in which rocketry played a central role. Yet he was not just a dreamer. He became one of the most active members of the Verein für Raumschiffahrt (Society for Space Travel), together with Willy Ley and Wernher von Braun – who would later develop the Saturn V rocket that took men to the Moon. His seminal investigations on interplanetary mission design appear in his book *Die Erreichbarkeit der Himmelskorper* (*The Attainability of Celestial Bodies*), published in 1925 (see Figure 8.7). The Hohmann transfer trajectories described in this work are still a valuable reference for any study on interplanetary travel. During the Second World War, however, he did not participate in the realisation of the German V1 and V2 rockets.

Hohmann died on 11 March 1945 during an allied bombing raid. In a comprehensive article by William McLaughlin, entitled 'Walter Hohmann's Roads in Space' it is said: 'Hohmann's great contribution to astronautical progress was the discovery of a new use for an old object: the ellipse.'

FIGURE 8.7. Walter Hohmann and the cover of his book on the accessibility of celestial bodies.

target – the length of the 'waiting time' (until the earliest occurrence of a favourable geometry to return to Earth) and the round-trip total duration. This last quantity deeply influences the possibility of manned exploration missions because of the need for carrying not only the fuel for manoeuvring a spaceship, but also the food needed to sustain a crew for extended periods of time. The present estimate is of about 5 kg of metabolic products (food, water and air) per person per day. A simple computation using the data in Table 8.1 yields a dramatic increase of the mass as the transfer time increases, and at present it can be considered one of the major difficulties in pursuing long-duration manned missions in interplanetary space. A technological breakthrough on space propulsion system is needed to fully realise Hohmann's dream.

GOING DEEP SPACE

The Voyagers' 'grand tour' of the outer planets was a fundamental step in the history of the exploration of the Solar System – a masterpiece of refined trajectory computation and engineering skills (Figure 8.8). It began when it was realised that an exceptional alignment of the outer planets Jupiter, Saturn, Uranus and Neptune – occurring every 175 years – could be exploited for

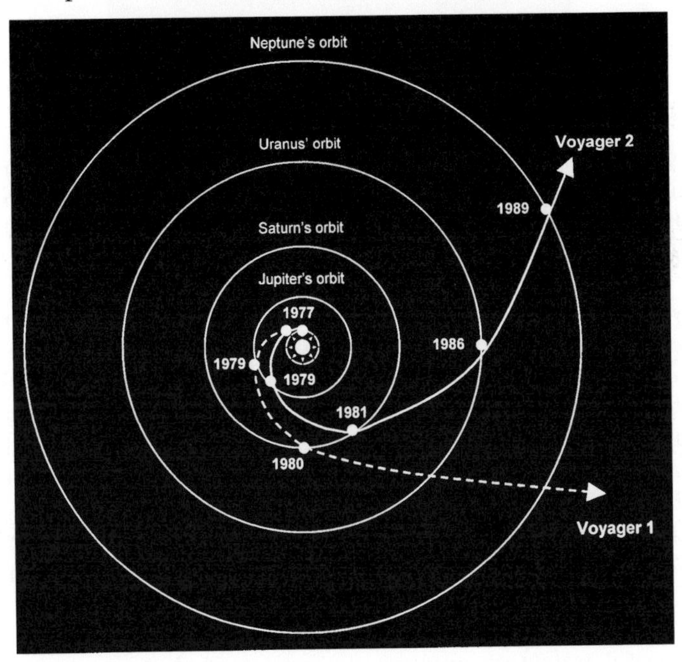

FIGURE 8.8. The trajectories of the Voyager spacecraft. The year in which planetary flybys occurred are marked. Only Voyager 2 visited all four outer planets. The scientific requirement of performing a very close passage past Saturn's largest moon, Titan, prevented Voyager 1 from continuing its journey to Uranus and Neptune.

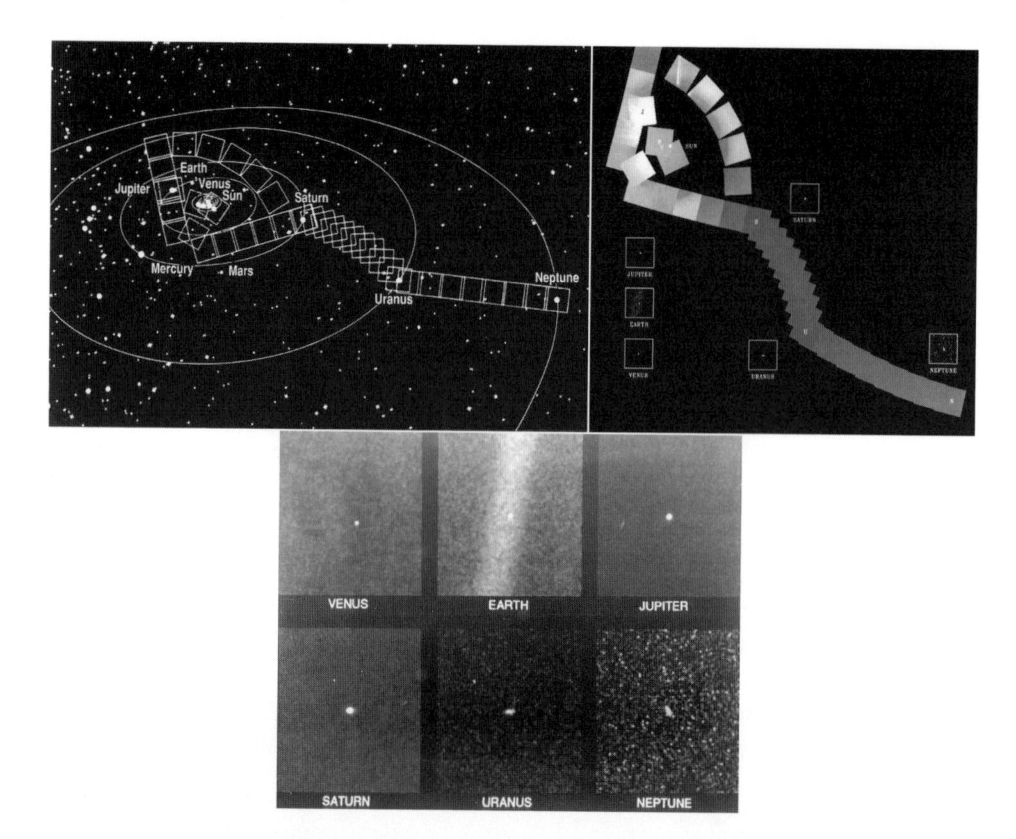

FIGURE 8.9. The first 'family portrait' of the Solar System. The Voyager imaging system was carefully pointed to take pictures covering the locations of the planets. (Top left) the strategy used for taking each frame; (top right) the actual images; (bottom) enlarged images of some of the planets. (Courtesy NASA/JPL.)

successively visiting these four planets (Figure 8.9). It was a unique opportunity because, as shown in the Hohmann accessibility table, the minimum energy transfer times to these planets are rather long – reaching, in the case of Neptune, up to 30 years. Planning four dedicated missions, each one directed to a giant gaseous planet, was unrealistic both on financial grounds and on the human timescale, as 30 years is much too large a fraction of a professional career, whether scientific, technical or managerial.

The possibility of using the gravitational pull of a planet's mass for increasing the size of an orbit without the need to use onboard propulsion (a 'gravity assist') had just been demonstrated by the Mariner 10 and Pioneer 11 spacecraft, which flew past Venus and Jupiter respectively, using their attraction as a velocity amplifier. Careful trajectory analyses demonstrated that due to the planetary alignment it was possible to allow a spacecraft to 'bounce' from one planet to the next – each time being accelerated to the point of successfully completing the tour in only 12 years. This was an opportunity not to be missed.

The twin Voyager 1 and Voyager 2 spacecraft were launched in summer 1977, and performed an astounding series of close fly-bys of the outer planets, approaching deep inside their satellite systems. As a result of the powerful gravity assists, both spacecraft achieved the Solar System escape velocity and are now travelling in the extreme regions of our planetary system. The Voyagers are still in good condition, and as a last present to us they have sent a snapshot of the Sun and the planets together, for the first time, as seen from outside the Solar System.

The art and science of gravity assist

The very first gravity assist was performed by Mariner 10 on 5 February 1974, when it sped past Venus using the planet's gravity to properly readdress its trajectory toward Mercury (Figure 8.10). The technique has quickly become an unavoidable choice for mission analysts because of the large amount of fuel savings, to the point that it is often referred to as the most efficient known form of space propulsion. But is there a man who deserves credit for such a brilliant idea? This is not a question that can be easily answered.

The idea of obtaining a free ride from gravity was already in the air during the pioneering years of space exploration, especially at the Jet Propulsion Laboratory – the organisation that manages interplanetary missions on behalf of NASA. After all, comets were long known to undergo significant orbital changes after close encounters with Jupiter, and by 1954 the British mathematician Derek Lawden had written about the benefits that could be derived from the exploitation of planetary masses.

In 1961 a young graduate student, Michael Minovitch, fascinated by the US planetary programme and working at JPL in a trajectory group, began

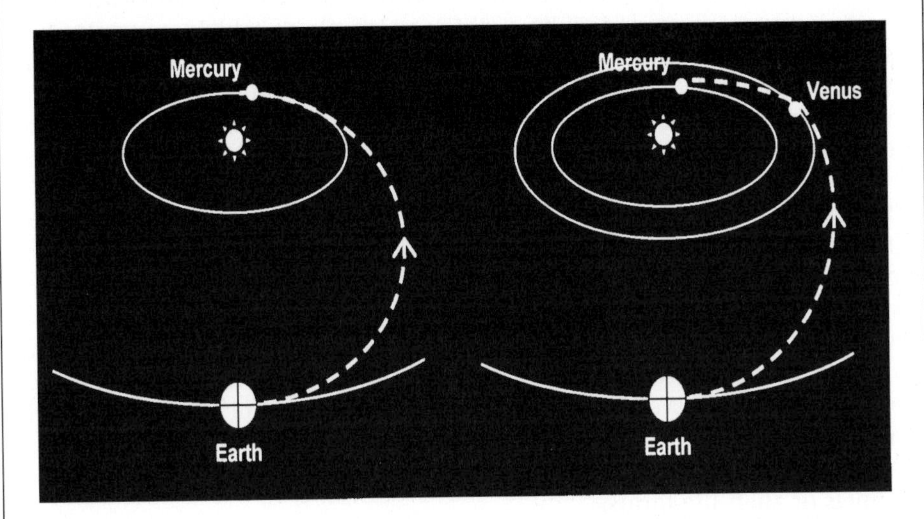

FIGURE 8.10. Mariner 10's gravity assist manoeuvre at Venus saved fuel and money. As a consequence of the significant decrease of the spacecraft's mass, a less powerful and therefore less expensive launcher was used.

investigating multiple planetary encounters. At university he had studied X-ray diffraction of crystals, which had no relationship with celestial mechanics; but, as often happens, this resulted in his approaching the problem of interplanetary trajectories with a clean state of mind. Instead of considering planetary perturbations as an annoying side-effect complicating the equations of motion, he instead thought of using them.

The same year Minovitch published a paper on the possibility of gravity-assisted manoeuvres. However, in an operational environment such as JPL, where missions in progress had to be kept under safe control, it did not attract the attention that it deserved. Nevertheless, some years later, upon seeing the 70% savings in fuel mass when applied to the Mariner 10 profile, a growing number of mission analysts started to work on the basic idea of gravity assist, succeeding in overcoming some practical difficulties, and demonstrated its feasibility.

Suspecting that some of the analysts at JPL were considering the discovery of gravity assist as the result of team work, Minovitch went as far as threatening to sue JPL for not recognising his concept. The riddle continued for many years, until in 1989 Minovitch was eventually invited to JPL as a special guest to witness first-hand the last Voyager encounter with Neptune.

Long before the Voyagers lifted off, the terrestrial planets Mercury, Venus and Mars had already been visited. Due to their proximity to the Earth they represented, in the 1960s and 1970s, the ideal targets for performing quick fly-bys to provide a first realistic assessment of the longstanding problem of life in the Solar System. Unfortunately the results frustrated the many expectations which had been steadily growing during the previous century. Mercury was discovered to be a hot and airless world, with a surface battered with impact craters. Venus is surrounded by a thick atmosphere which prevents direct observation of the surface. For a long time all that was known from the Russian attempts to land on Venus was the extremely high values of the temperature (about 500°) and pressure (90 bars) at ground level, and the acid corrosive chemical composition of the atmosphere, which allowed the probes to survive for only a few hours before major systems failure.

Several US and Russian missions did not succeed in approaching close to Mars, but the planet was eventually reached in 1964. The Earth's brother planet was on the front line as far as extraterrestrial life was concerned. In the nineteenth century the Italian astronomer Giovanni Schiaparelli (1835–1910) claimed to have observed an intricate network of channels on the surface of Mars (Figure 8.11). His findings were promoted around the world; but the Italian word *canali*, meaning channels, was mistranslated as *canals*. Many people therefore assumed that they were artificial, with obvious implications. To the cameras onboard Mariner 4, however, a lunar-like landscape appeared, with no signs of canals, channels or alien life-forms.

Attention was then turned to the small bodies of the Solar System, and in particular to the 1986 return of Halley's comet. From the point of view of celestial mechanics it was a challenging target, because its orbit is highly

FIGURE 8.11. Giovanni Schiaparelli and the martian channels.

eccentric and, moreover, retrograde (opposite to the planets' direct motion). A spacecraft leaving the Earth inherits the orbital velocity of our planet – almost 30 km/s – and if a retrograde orbit is required, an unrealistic 60 km/s ΔV budget should be taken into account. Half of this is needed for cancelling the initial velocity of the spacecraft, and the other half for initiating retrograde motion. The reason for avoiding an object on a retrograde orbit when the spacecraft motion is direct is that the two bodies approach from opposite directions and at encounter their velocities add up. In the case of comet Halley this translates into 70 km/s relative velocity. Exploiting Jupiter's mass for reversing the spacecraft motion was an option, but the resulting mission profile was much too complicated and long-lasting. The possibility was therefore discarded, and the risky high-velocity fly-by was the only chance remaining.

A small fleet of spacecraft was then prepared: two Vega spacecraft from Russia, the tiny Japanese probes Sakigake and Suisei, and the first European interplanetary mission, named Giotto after the celebrated Italian painter who depicted the 1301 return of comet Halley. The Russian and Japanese missions arrived first, but they were badly damaged by collisions, and the few images that they sent back to Earth could not be clearly interpreted. Yet their flight was not useless. Giotto exploited the Vega trajectory data for heading directly to the nucleus hidden inside the dense bright coma of the comet. On 13 March 1986, the 'night of the comet' at the European Space Operations Centre, in Darmstadt, Germany became an exciting historical event. The major threat to the success of the mission was associated with a small mirror, used for imaging the comet, protruding out of the shielded body of the spacecraft. Every impact of cometary dust would result in a small crater on the polished mirror's surface, leading to a reduction in reflectivity. Would the mirror survive until the crucial phase of the mission? At the incredible speed with which Giotto was flying towards the comet, it was a matter of seconds. One minute before closest approach, the spacecraft was 4,000 km away – and 30 seconds later the distance was halved. Much to the happiness of scientists and engineers, Giotto succeeded in performing a close fly-by of the nucleus, approaching it to within 600 km. The onboard camera survived long enough to send clear images of a $16 \times 8 \times 8$ km irregular body with powerful jets of gas and dust emanating from cracks in its surface.

HIGHWAYS TO THE PLANETS

At the beginning of the 1990s all the Solar System planets had been closely imaged by interplanetary probes, mostly during quick fly-bys. The next step for planetary exploration and the new challenge for spaceflight dynamics was to deploy large orbiting spacecraft, allowing observations extended in time and, depending upon the physical nature of the target, surface landers and rovers or atmospheric probes. Sample return missions – especially to asteroids and comets – also began to represent a feasible goal.

These scenarios imply basic rendezvous mission requirements as far as the overall ΔV is concerned. As deduced by looking at Table 8.1, in which plain Hohmann strategies are shown, Mars and Venus are relatively accessible, thus posing no particular problems for extensive exploration. In 1992 the Magellan spacecraft successfully entered into orbit around Venus, completing the first mapping of its surface by using onboard radar for piercing the dense layers of clouds surrounding the planet. It has recently been followed by the European Venus Express mission.

Mars is the most visited among the planets. After the 1976 twin Viking missions which both orbited and landed, the planet is now permanently kept under close observation. Among the many successful missions launched in the last decade it is worth mentioning NASA's Spirit and Opportunity rovers, which have been roaming over the martian desert terrains, and ESA's Mars Express, carrying powerful deep-penetrating radar for detecting the huge water ice reservoirs that are thought to be hiding just below the surface.

When Mercury and the outer planets are taken into consideration an extensive use of the gravity assist technique becomes mandatory, because no launch system could provide the required energy. An innovative trajectory design was then needed for the natural follow-up of the Voyager missions: sending spacecraft into orbit around Jupiter and Saturn.

Relying on past experience and success, mission analysts grew confident in planning multiple gravity assists using the Earth and Venus for achieving transfer trajectories able to reach those distant targets. This is how in December 1995 the Galileo spacecraft managed to eventually perform the planned orbit insertion manoeuvre at Jupiter, thus becoming the first artificial satellite of the giant planet. Shortly afterwards, in October 1997, the Cassini mission to Saturn began its long journey, carrying with it ESA's Huygens probe. Huygens' descent into the atmosphere of the large satellite Titan in January 2005 is one of the major achievements of space science.

Gravity-assisted trajectories can be considered the space analogue of terrestrial

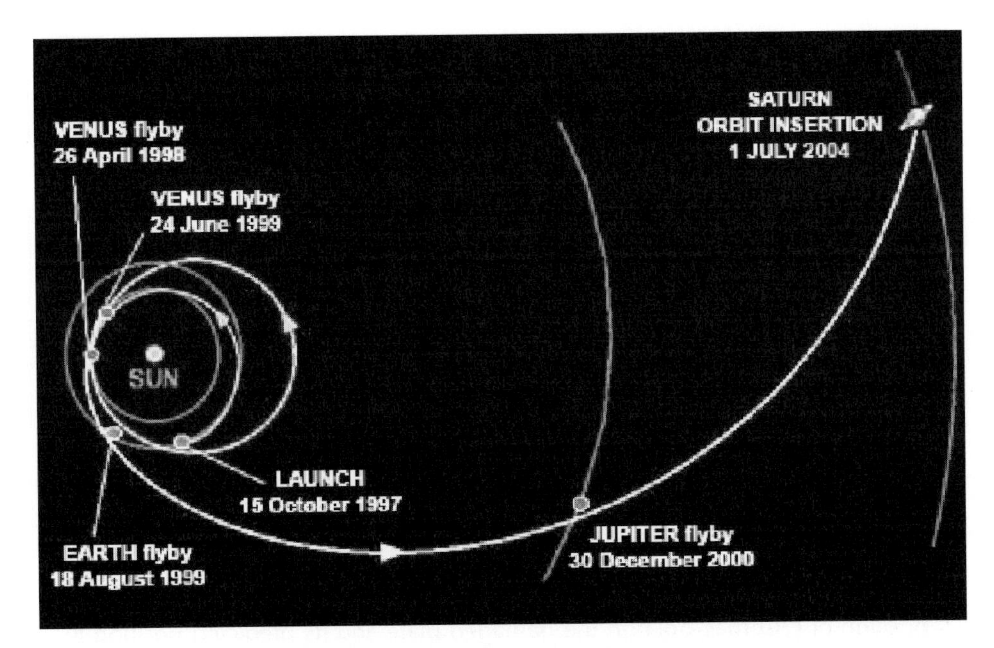

FIGURE 8.12. The trajectory of the Cassini spacecraft.

highways, in that they allow safe connections among long-distance targets. These celestial routes have also been named accordingly: Galileo was on a VEEGA (Venus–Earth–Earth Gravity Assist) track, while for Cassini an Earth–Venus–Venus–Jupiter (EVVJGA) has been chosen (Figure 8.12).

As in 'cosmic billiards', targeting planets has quickly become a good game to play. One of the largest near-Earth asteroids, Eros, was reached by the NEAR spacecraft in 2001, after a polar encounter with the Earth provided the significant change in inclination (10°) required to match the orbital plane of the spacecraft with that of the asteroid – a basic requirement for a rendezvous mission profile.

The current ESA Rosetta cometary mission appears even more challenging. Three Earth gravity assists and one Mars fly-by will take it to the correct orbital path for slowly approaching comet Churyumov–Gerasimenko and delivering a small lander on its icy surface. This will allow *in situ* sampling of the pristine material composing the nucleus, and following in great detail the onset of activity as the comet approaches the Sun.

LAST BUT NOT LEAST

From the beginning the space age proceeded astoundingly fast. Only two years after the first artificial satellite, Sputnik, was launched into low Earth orbit, a spacecraft flew close to the Moon, 1,000 times further away. Two years later the first interplanetary probe escaped the attraction of the Earth and began orbiting

A comet in the Garden of Eden

Rosetta's target – comet 67P/Churyumov–Gerasimenko – tells a rather peculiar and lucky tale.

Firstly, it was not discovered in the usual way. Comet chasers spend long nights – often in the cold – scanning the celestial sphere with wide-field binoculars, knowing by heart entire regions of the sky in order to recognise the faintest intruder. They also need to be quick enough to communicate a discovery to the Central Bureau for Astronomical Telegrams to have the privilege of naming the comet.

This was not the case for Klim Ivanovich Churyumov and Svetlana Ivanovna Gerasimenko, who in September 1969 were participating in a photographic survey of known periodic comets, carried out at the Alma-Ata Astronomical Observatory in Kazakhstan. Gerasimenko noticed that one of their plates, on which comet P/Comas–Solà was supposed to appear, was badly damaged. However, on closer investigation a comet-like object was still recognisable, and was therefore identified as that comet. One month later, while reducing the Alma-Ata observations, a 2-degree discrepancy was measured between the position of P/Comas–Solà on the damaged plate and its predicted position in the sky – which was too much to be accounted for by observational errors. The only possible explanation was the presence of another comet, which was eventually confirmed on other plates from the same survey. As Gerasimenko stated in a recent interview: 'The spoiled plate brought us good luck!' And enduring good luck, it should be added, as almost 35 years later the same comet again rewarded its discoverers by substituting for another target comet – P/Wirtanen, which was not reachable by Rosetta due to the 1-year postponement of the spacecraft's launch date.

This is not the only coincidence concerning comet Churyumov–Gerasimenko. According to local tradition, Alma-Ata is the place where the legendary apple tree of the Garden of Eden had grown. Thus a comet discovered from the cradle of mankind will hopefully help to unveil the birth of our Solar System. It could not have shared a different fate.

the Sun as a tiny artificial planet, and at the same time, in April 1961, Yuri Gagarin successfully completed three orbits around our planet. Eight years afterwards, men walked on the Moon.

Excitement about space exploration grew considerably, and the wish of extrapolating those incredible early years into the future led to overoptimistic scenarios of large infrastructures orbiting the Earth and lunar colonies by the year 2000. The basic reason why it is not so is the aforementioned lack of a technological breakthrough in space propulsion systems. Chemically-fuelled spacecraft have intrinsic limitations which spaceflight dynamics has tried to compensate by exploring novel trajectory designs for achieving targets otherwise out of reach. Over the years these celestial routes have been followed by highly sophisticated unmanned spacecraft, and almost all major Solar System bodies have been visited. To date there are only two important targets left, located at the 'opposite' sides of the Solar System: Mercury is too close to the Sun, while Pluto is too distant. But not for long. NASA's Mercury Messenger and ESA's BepiColombo missions will soon be orbiting Mercury, while in 2006 the New Horizons mission began a long journey that in 2015 will take a spacecraft to fly-by Pluto and, hopefully, a couple of transneptunian objects.

Slingshotting from planet to planet

During their tours of the giant planets in the late 1970s and 1980s, each of the two Voyager spacecraft used Jupiter's gravity to be hurled onwards to Saturn. This so-called 'gravity assist' caused the spacecraft to speed up, relative to the Sun, by about 16 kilometres per second or around 57,500 kilometres an hour. Yet gravitation is a two-way phenomenon, so Jupiter's motion was also affected by the encounter: indeed while the Voyagers were accelerated, the giant planet was slowed in its orbit around the Sun – but since it is incredibly massive relative to the spacecraft, the corresponding delay is hardly measurable, resulting in about one centimetre in 30,000 million years! Although Voyager 1 headed out of the Solar System following its Saturn encounter, further gravity-assist fly-bys of Saturn and Uranus enabled Voyager 2 to complete its tour of the four gas giants out to Neptune.

In all cases, extensive use of gravity assists by the planets is foreseen; but BepiColombo is powered by a new type of propulsion. Instead of the explosive burning of chemical compounds which leads to the sudden release of large quantities of energy in a very short time (impulsive thrust), electric propulsion acts more quietly and for longer timespans. It is based on accelerating electrically charged particles which are continuously ejected from the engine, thus providing continuous low thrust, which ultimately provides a higher overall ΔV with respect to high-thrust chemical engines.

Accounting for such a peculiar 'perturbation' is a completely new problem for celestial mechanics, and the finding of optimal low-thrust electric propulsion trajectories has opened an entirely new field of study. As an example, escape from the Earth's gravitational field may result in slow outward spiralling, while capture around another celestial body is obtained by a symmetrical inward spiralling trajectory. An example of using electric propulsion for an Earth–Moon transfer is shown in Figure 8.13, obtained by tracing the orbit of the spacecraft in a suitable reference frame, rotating with the same angular velocity of the Earth and the Moon.

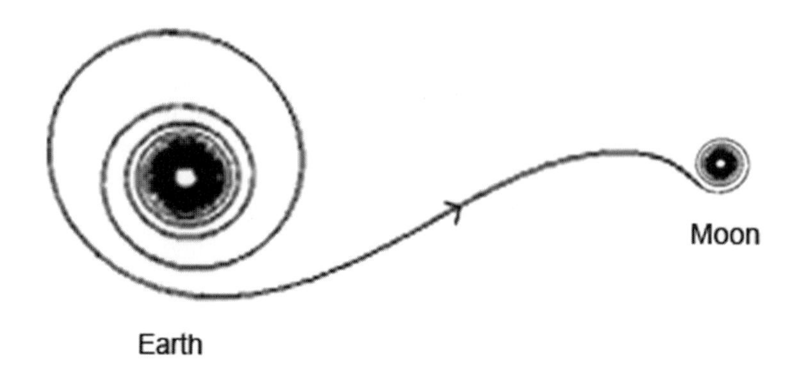

FIGURE 8.13. The trajectory of an electric propulsion-powered spacecraft leaving a 500-km-altitude circular orbit around the Earth, aimed at the Moon. A selenocentric 200-km-altitude circular orbit is reached after approximately three months.

The SMART-1 spacecraft, designed by the European Space Agency for testing electric propulsion in space, has recently reached the Moon on a similar trajectory.

9

Lords of the rings

The voyage of discovery is not in seeking new landscapes, but in having new eyes.

Marcel Proust

Four hundred years of observations, from Galileo Galilei to the Voyager and Cassini missions, allow comparative analysis of the ring systems surrounding the outer planets. Although their extension and composition significantly differs from planet to planet, celestial mechanics represents the common driving force for explaining their present appearance. Traditional and exotic commensurable motions among ring particles, satellites and moonlets are frequently found, with intriguing consequences on the stability and confinement of the rings. Their dynamics can be studied by means of three-, four- and N-body subsystems, thus representing the 'dream come true' for celestial mechanics.

RINGED WORLDS

When in 1610 Galileo turned his telescope toward Saturn he described his observations using the following anagram (a sort of 'copyright' protection of the time) to be solved in Latin: smaismrmilmepoetaleumibunenugttauiras. Thinking that Galileo had discovered two new martian moons, Kepler deciphered it as *Salve umbisteneum geminatum martia proles*. But he was wrong. The correct solution is *Altissimum planetam tergeminum observavi* – 'I have observed the highest planet as tri-form.' The 'highest planet' is Saturn, because it was the most distant planet known at the time (Uranus was not to be discovered until 1781), while the 'tri-form' nature refers to the planet appearing to be accompanied by two other bodies. The obvious conclusion that Saturn had large satellites orbiting around it, similar to what Galileo had just discovered in the case of Jupiter, was not confirmed by his observations performed in 1612, when the planet again appeared to be alone in the sky.

Over the ensuing years the joint efforts of skilled astronomers and brilliant minds revealed that Saturn is surrounded by a dense, flat ring of particles, and that their periodic disappearance is due to the change of perspective with respect to an observer on Earth (Figure 9.1). In 1675 Giovanni Domenico Cassini first

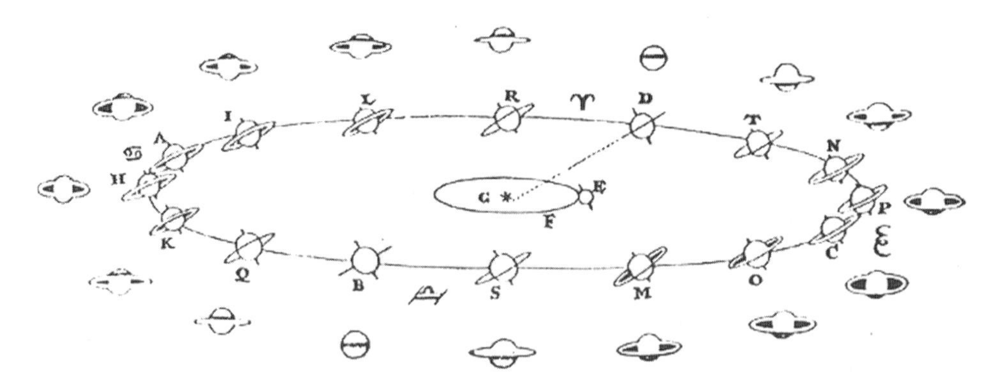

FIGURE 9.1. Because of the non-zero obliquity of Saturn, the rings undergo 'seasonal' effects as the planet revolves around the Sun. In particular, their orientation changes with respect to an observer on Earth, by which the appearance of the planet in the field of view of a telescope is observed at the different positions of Saturn along its orbit.

observed that the rings exhibited structure, and that a large dark gap separates the outer A ring from the brighter inner B ring. Almost a century later, Immanuel Kant was the first to propose that the rings' structure is not solid, but is composed of small particles. In 1789 Laplace studied the stability of a solid ring and was led to share the conjecture of many thin ringlets. Finally, in 1859 James Clerk Maxwell established that the rings are composed of swarms of particles in orbital motion, and also proved their stability. Ring dynamics was born, and it is reported that the Astronomer Royal, George Biddell Airy, exclaimed: 'It is one of the most remarkable applications of mathematics to physics that I have ever seen.'

As observational techniques improved, more divisions within the rings were found, and their connection with the motion of the nearby satellites, Mimas, Enceladus, Tethys and Dione, slowly emerged. The development of sophisticated theories for explaining satellite–ring interactions has quickly become one of the most advanced and successful fields of celestial mechanics.

Today the rings of Saturn can be easily seen with a small telescope, and anyone can feel the same sense of astonishment experienced by those early observers. The spectacular images sent back by the Voyagers during their 1980–81 fly-bys of Saturn and by the current Cassini mission, also allow us to distinguish, at first glance and in amazing detail, the complex structure of the ring system (Figure 9.2).

Saturn, however, although retaining a position of excellence, is not the only ringed planet. Within a decade, starting in 1977, ground-based and spacecraft observations have shown that *all* the outer gas giants have rings. Uranus is surrounded by a remarkable system composed of several narrow rings and moonlets (Figure 9.3), and an unusually high eccentricity characterises them as unique among the ring systems. Neptune has three major rings (Figure 9.4), one of which is characterised by discontinuous arcs – a phenomenon that was investigated for more than a decade before a convincing explanation was found.

A strange family

The four outer planets take their names from the Greek and Roman gods. Uranus was the son of Mother Earth, generated by her upon emerging from chaos at the beginning of time. He, in turn, became father of the Titans, the youngest of them being Chronos – the bringer of time – later identified with the Roman deity Saturn. As an act of rebellion, Chronos killed Uranus by castrating him, banned his brothers, and attained complete power. He then married his sister Rhea, but under the prophesy that he would be dethroned by one of his sons, Chronos began devouring them as soon as they were born. Rhea succeeded in saving his third child Zeus (Jupiter) by hiding him in a cave on the island of Crete, where he was nourished by the goat Amalthea. Once Zeus had grown, he planned his revenge. Disguised as a servant he made Chronos drunk to the point of vomiting his brothers still alive, including Poseidon (Neptune). With them he made war on Chronos, and after victory he locked his father in the underworld, where he was to remain forever.

When Galileo found that the two supposed 'satellites' of Saturn had mysteriously disappeared, he asked: 'Has Saturn again devoured his own children?'

FIGURE 9.2. The fine structure of Saturn's rings and the radial spokes, as imaged by Voyager 1. The large dark gap is the Cassini division, separating the A ring (outer) from the B ring, where the spokes appear. (Courtesy NASA/JPL.)

FIGURE 9.3. Uranus's rings are thin and narrow. The brightest is the ε ring (left), imaged by Voyager 2 during its encounter with Uranus in 1986. (Courtesy NASA/JPL.)

FIGURE 9.4. A Voyager 2 image of the two outer narrow rings and the inner diffuse halo that form the bulk of Neptune's ring system. (Courtesy NASA/JPL.)

FIGURE 9.5. Jupiter's faint main ring, imaged by the Galileo spacecraft in eclipse while passing on the dark side of the planet (top), and a close-up view showing its fine structure. (Courtesy NASA/JPL.)

Jupiter exhibits only faint dusty rings (Figure 9.5); but again, determining their origin, starting from dynamical considerations, has proved challenging.

As a consequence, the phenomenology of planetary rings has quickly become particularly rich. Large-scale structures are usually shaped by the motion of large nearby satellites through the combined action of different types of orbital resonance. The many peculiar features observed are studied on a case-by-case basis, and often involve subtle gravitational and non-gravitational perturbations.

The ring particles also exhibit substantial differences. The main component of Saturn's bright rings is water ice, while the particles range in size from micrometric dust to large boulders tens of metres wide. Rocky material is also present, especially in the much darker ring systems around Uranus and Neptune. Jupiter's tenuous dusty ring system is short-lived, and a source for its replenishment must be found, possibly among the small satellites of the planet.

The intrinsic differences in the ring and satellite systems of the outer planets is shown in Figure 9.6. This diversity raises a number of questions involving different fields of science, and represents a test-bench for many theories on the origin and evolution of many-body systems ranging from global dynamics to statistical approaches.

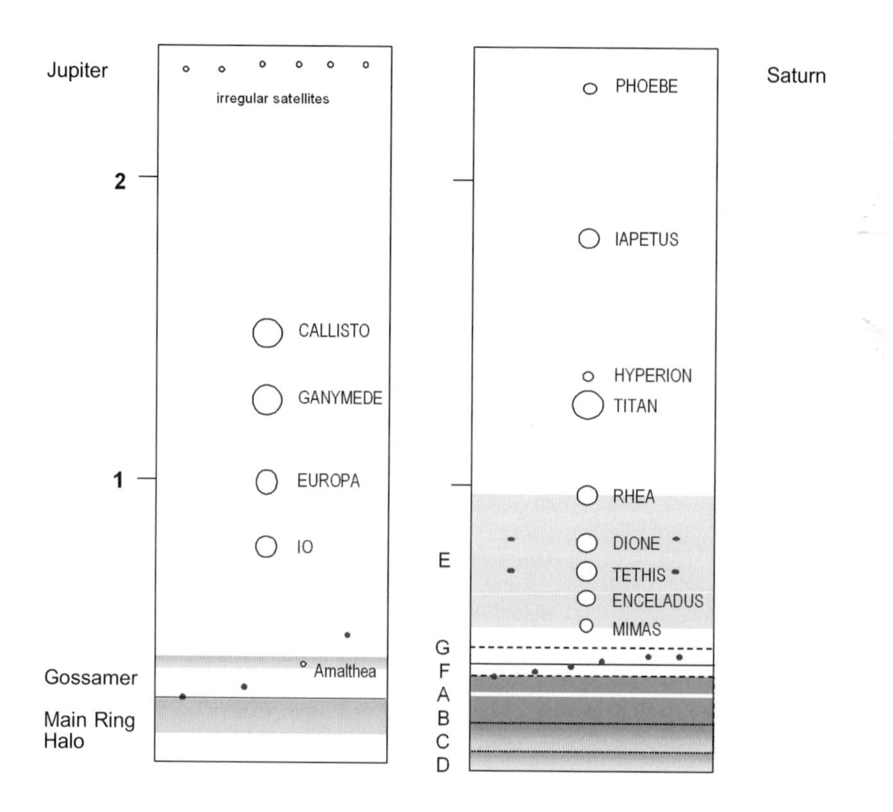

FIGURE 9.6. (above and on facing page) A graphical representation of the complex systems of rings and satellites of the outer planets. For comparison, distances are computed in units of the radius of the planet and using a logarithmic scale. The bottom of each diagram therefore corresponds to the cloud top, while unity marks a distance equal to the radius of the corresponding planet, number 2 a distance ten times larger, and so on. Open circles indicate satellites discovered by ground-based observations, while full circles indicate the small satellites and moonlets discovered from space. The size of the circles approximately indicates the size of the corresponding celestial bodies. Only some of the many moonlets and irregular satellites are shown in the diagram as representative of their location. Rings are indicated by lines when they are narrow and by grey regions when their extension is significant.

FORBIDDEN REGIONS

In 1850 the French astronomer Edouard Roche (1820–1883) developed a mathematical theory for computing the boundary of a particularly dangerous region around a celestial body. When applied to the planetary case, the *Roche limit* defines the distance below which any self-gravitating body (an object kept together only by gravitation) can be disrupted by the strong tidal forces exerted by the planet. This is possibly what happens to comets that undergo an

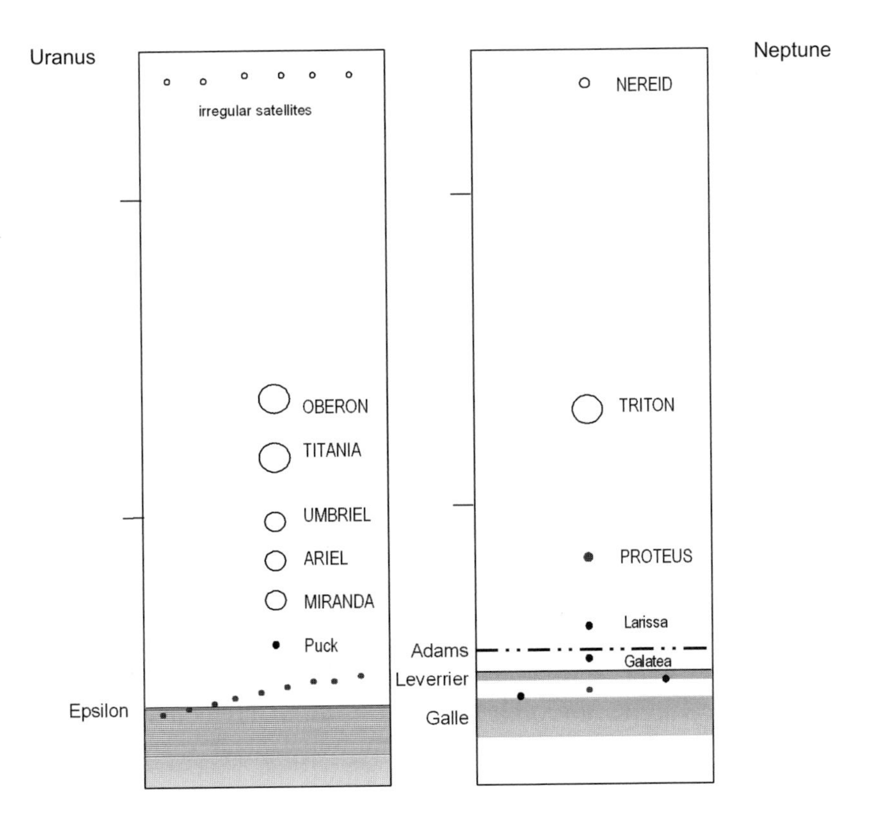

FIGURE 9.6 (continued)

Proceed at your own risk

The Roche limit R_L is computed as a function of the radius of the planet R_P and the densities of both the planet ρ_P and the celestial body ρ moving around it:

$$R_L = (2\ \rho_P / \rho)^{1/3}\ R_P.$$

Assuming the two reference densities of 3 and 0.5 g/cm^3 as representative of rocky bodies (such as asteroids) and icy bodies (such as comets), it is possible to compute the corresponding Roche limits for all planets. In the case of Jupiter the dangerous region is located at 129,000–235,000 km; for Saturn, 86,000–157,000 km; for Uranus, 47,000–85,000 km; and for Neptune, 49,000–89,000 km.

Note that celestial bodies can be (and have been) found inside the Roche limit, provided that forces other than gravitation keep them together. As an example, with the above computations applied for the Earth, the corresponding Roche limits are found at 19,000 km and 34,000 km. The existence of artificial satellites orbiting at distances well below these values (for example, the LEO satellites, located at altitudes of a few hundred kilometres) is justified by their having sufficient tensile strength to counterbalance the tidal force induced by our planet.

extremely close encounter with Jupiter – such as Shoemaker–Levy 9 (see Chapter 5).

The Roche limit is very useful when studying planetary rings, as it identifies the minimum distance from the planet where material could coalesce to form natural satellites instead of remaining as dispersed dust.

In reviewing the summary of the physical characteristics of the natural satellites of the outer planets, it is possible to see that they all lie outside the Roche limit and that their size decreases on approaching it. Jupiter's moons Adrastea and Metis orbit near the Roche limit, and within Saturn's system the closest satellite is Mimas (with a semimajor axis of 185,600 km). Inside the 'forbidden region' only rings and tiny satellites (moonlets) are found (Figure 9.6).

Another peculiar dynamical configuration – useful for understanding the structure of planetary rings – is the *corotation orbit,* with a semimajor axis that corresponds to a period of revolution with the same value as the rotation period (the length of the day) of the planet. Any corotating particle therefore remains stationary as seen from the planet below, thus representing the planetary equivalent of the geostationary ring used for telecommunication satellites located some 36,000 km above the Earth's surface (see Chapter 4). In practice, knowing, for example, that Saturn completes a full rotation in 10.66 hours, and using this value in Kepler's third law, it is possible to calculate that the radius of the corresponding corotation orbit amounts to about 112,000 km – well inside the B ring. Repeating the procedure for all the outer planets it is interesting to note that the corotation radii of Jupiter and Saturn stay within the Roche limit, while for Uranus and Neptune they fall outside the limit.

The Voyager images show the presence of radial dark structures, called *spokes* (Figure 9.2), along the corotation region within Saturn's B ring, persisting over relatively short timespans. They are probably related to the action of the magnetic field of Saturn on the small charged particles that shape the spokes.

JOVIAN HALOS

An unexpected discovery of the Voyager missions was the detection of a faint ring system around Jupiter; and the Galileo mission, which orbited Jupiter during the mid-1990s, has allowed us to study its fine structure in detail. A main ring (Figure 9.5) is surrounded by two diffuse halos – one extending outward and the other inward. The main ring, spanning more than 6,000 km, is composed of microscopic particles of rocky nature, the motion of which is strongly influenced by Jupiter's magnetic field. Two small satellites – Adrastea and Metis – orbit very close to its outer border, which roughly coincides with the Roche limit.

Within the outer halo two narrow regions can be identified: the inner and outer Gossamer rings at 129,000–182,000 km and 182,000–225,000 km. The orbits of the two small moons Amalthea and Thebe are inside this region, thus justifying a possible collisional explanation of the formation of Jupiter's ring system as the result of meteoroidal impacts on the small moons and the consequent ejection of large quantities of dust around the planet.

SIGHTSEEING SATURN

The ring system of Saturn has seven components, labelled A to G (following the sequence of their discovery), and separated by divisions of different widths – regions apparently devoid of particles or characterised by an abrupt change in size, density and arrangement. In particular, the Cassini division – extending from 117,580 km to 122,170 km from the centre of the planet and marking the outer edge of the B ring – is connected with the motion of the satellite Mimas. By computing the ratio between the orbital period of Mimas and that of a particle located at the inner edge of the Cassini division, the value 0.504 is found – which is remarkably close to the fraction ½. Other major divisions corresponding to commensurable motions are Encke's division, in a 5:3 resonance with Mimas, and the outer edge of the A ring (Keeler's division) in a 7:6 orbital resonance with Janus.

The tiny moonlets orbiting among the rings also play a deciding role in shaping the overall structure of the system. Narrow rings are kept together by the action of the shepherd satellites orbiting on both sides of the ring, the gravitational force of which provides the necessary confinement of the particles of the ring. This is the case for Prometheus and Pandora – the inner and outer shepherding satellites of the F ring. The co-orbitals Janus and Epimetheus (whose peculiar interaction is discussed in Chapter 2) bridge the gap between the F ring and the G ring.

Saturn's rings are designated with capital letters, in the sequence D, C, B, A, F, G and E, from the inner to the outer (see Figure 9.6 and Table 9.1). The D ring is the innermost and is extremely faint. The brightness of the disk increases

Table 9.1. The main features of Saturn's rings, including gaps and shepherding satellites.

	Distance (km)		Distance (km)
Saturn equator	60,268	Keeler division	136,530
D inner edge	66,900	A outer edge	136,775
D outer edge	74,510	Atlas	137,700
C inner edge	74,568	Prometheus	139,400
Maxwell gap	87,491	F ring centre	140,180
C outer edge	92,000	Pandora	141,700
B inner edge	92,000	Epimetheus	151,400
B outer edge	117,580	Janus	151,500
Cassini division		G inner edge	170,000
A inner edge	122,170	G outer edge	175,000
Encke division	133,589	E inner edge	181,000
Pan	133,600	E outer edge	483,000

Ring data are from the Cassini–Huygens mission (courtesy ESA)

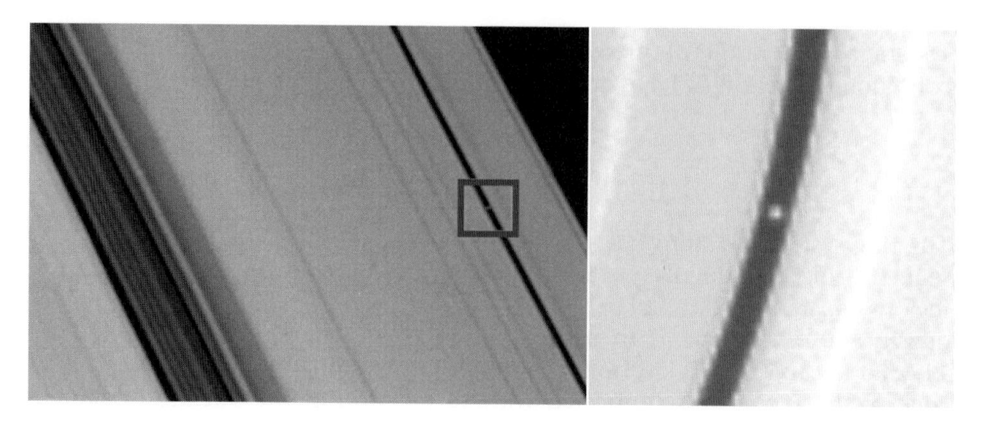

FIGURE 9.7. (Left) the discovery image of Pan, the position of which is marked by the square. Tiny ringlets can be seen inside the Cassini division. (Right) a recent image of the satellite acquired by the Cassini mission. (Courtesy NASA/JPL.)

Play it again, Pan

In 1990 – almost ten years after the historical Voyager encounters with Saturn – a new satellite was discovered on the images returned by the spacecraft. The reason why it took so long to spot is that the tiny new moon is only 10 km across and hides inside Encke's division (Figure 9.7). Its presence was revealed by the American astronomer Mark R. Showalter, as predicted by studying the gravitational perturbations exerted on the borders of the neighbouring rings, which produce characteristic wave-like patterns.

The moonlet keeps the division clear, thus giving a striking example of the tight relationship between the macroscopic structure of the rings and the orbital motion of moonlets. This eighteenth satellite of Saturn has been named after the Greek playful musician god Pan.

outward, through the C ring and especially through the B and A rings, which are formed of myriads of ringlets. The F and G rings are very dim and are difficult to observe. The E ring resembles an extended dusty halo embracing the orbits of some of the inner satellites of Saturn, which are thought to be the source of the particles forming the ring - in particular, Enceladus.

Although the overall radial structure of Saturn's rings extends over several hundred thousand kilometres from the planet's centre, they rarely exceed 1 km in thickness.

ELLIPTIC RINGS

The rings encircling Uranus were first detected by means of a stellar occultation - an observational technique that takes advantage of the passage of a planet in front of a bright star, as seen from an observer on the surface of the Earth. Occultations have been traditionally used for probing the atmosphere of distant planets, because the star does not disappear instantaneously; rather, the luminosity decreases, following the thickness of the planet's atmospheric layers. A similar effect is observed when the light of the star encounters the rings and gaps surrounding a planet, thus allowing us to estimate their location, their extension and the density of the particles in the rings.

In 1977 this method was applied during observations of the occultation of the star SAO 158687 by Uranus. The observers expected typical oscillations in the luminosity of the star as Uranus passed in front of it; but it faded and brightened several times before and after the occultation. Further observations and detailed analysis of data revealed the existence of a system composed of several narrow rings around Uranus. During the occultation the star had been obscured by the rings and restored to brightness when appearing through the divisions.

Direct confirmation came from the Voyager 2 spacecraft during its encounter with the planet in 1986: eleven rings of significant eccentricity, and moderate inclination with respect to the planet's equator. These close-range observations also led to the discovery of several moonlets within the ring system, thus confirming their fundamental role in shaping its structure. Among them, Cordelia and Ophelia have been recognised as shepherd satellites of the outermost ring of Uranus. More recently two new outer rings were discovered by the Hubble Space Telescope.

Uranus's rings are denoted with Greek letters, from the α ring to the ε ring. These names (unlike for Saturn) indicate increasing distance from the planet.

ARCS IN THE SKY

Neptune's destiny seems to be of always raising lively debates in the astronomical community. The events leading to the observation of the rings was as peculiar as the circumstances of the planet's discovery (see Chapter 1). By the mid-1980s, ground-based observations appeared to be consistent with the presence around Neptune of a faint, narrow and incomplete ring - a puzzling hypothesis for celestial mechanics.

The riddle was eventually solved in 1989, thanks to the skilled Voyager 2 flight dynamics team. A difficult manoeuvre was planned, consisting of turning the spacecraft imaging system back to the planet while it passed inside Neptune's shadow in order to avoid being blinded by the brightness of the planet and to detect any faint ring system. The strategy was successful, and more than 800 images of Neptune's rings were returned to Earth. These images clearly show that the planet is surrounded by a ring system composed of at least by two narrow rings and a larger ring diffusing towards Neptune's cloud-tops (Figure 9.4). They have been named Adams, Leverrier and Galle, after the three astronomers responsible for the discovery of Neptune.

The ring afterwards named Adams was responsible for the puzzling ground-

A puzzling result

While the Voyager 2 spacecraft was still far from Neptune, the best image of the planet available, obtained by the most powerful ground telescopes, showed only a small bright disk on a dark background. Therefore the only way to determine whether the last giant planet also had a ring system was to repeat the stellar occultation experiment which had proved successful in the case of Uranus.

In May 1981 the first opportunity of this kind was exploited by a number of teams all over the world; but results were not encouraging. In 1983 another occultation appeared particularly favourable because it was observable by the large telescopes on Mauna Kea, Hawaii. Yet no clear indication was obtained to conclude that if rings were present around Neptune they must have been very narrow – no larger than 300 km.

When a new opportunity arose the following year, only two teams – one led by the Paris Observatory and the other by the Inter-American Observatory – were willing to try. This time the luminosity of the reference star dropped clearly at a distance of about three Neptune radii, indicating the presence of an extremely narrow ring, possibly less than 15 km wide. Unfortunately the discovery had a puzzling anomaly. No symmetrical decrease in the intensity of the light was recorded when the star emerged on the other side of the planet, as if the ring was not complete. Should it be considered an inaccurate observation, or was celestial mechanics faced with a new problem – the existence of ring arcs?

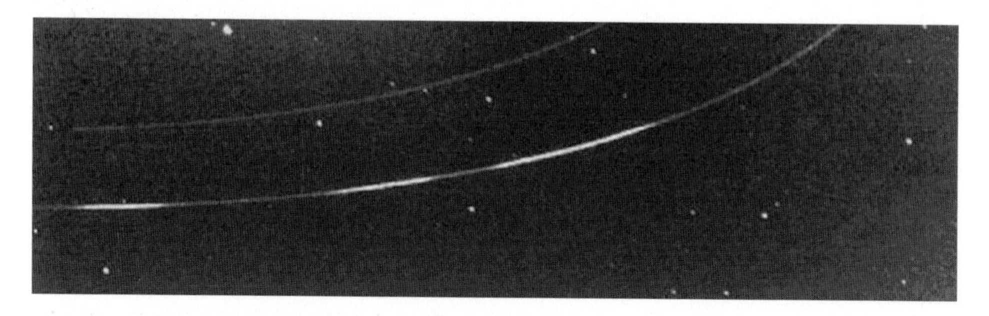

FIGURE 9.8. The three bright arcs in Neptune's Adams ring, imaged by Voyager 2. (Courtesy NASA/JPL.)

based observations, because three brighter regions clearly appeared along the ring (Figure 9.8), indicating that the supposed arc was due to particles trapped inside well-defined regions of space. Also found was the expected small crowd of moonlets associated with the ring system. These were later named Naiad, Thalassa, Despina, Galatea and Larissa (Figure 9.6).

The experience gained in investigating the complex dynamics of Saturn's ring system proved to be extremely helpful in trying to account for the strange behaviour of the Adams ring. In particular, attention was drawn to Galatea – a 150-km satellite orbiting just inside the Adams ring – but more importantly in a 42:43 mean motion resonance with it. This was the basic requirement for supporting a subtle mechanism called *corotation inclined resonance* (originally developed by the American astronomer Peter Goldreich), which could account for the existence of relatively stable and isolated regions within the same ring (Figure 9.9). To mark the bicentenary of the French revolution, the three bright arcs were named Liberté, Egalité and Fraternité.

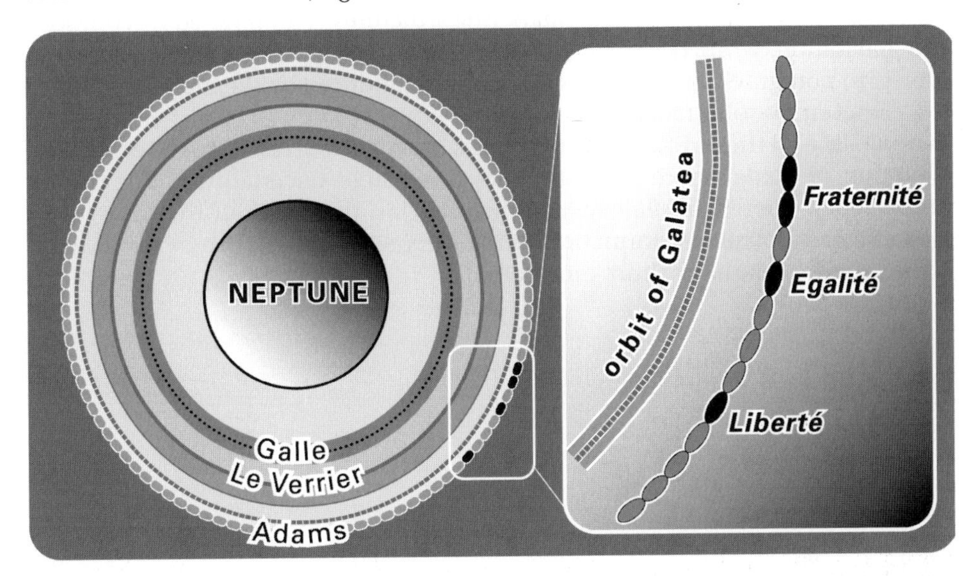

FIGURE 9.9. The predicted corotation regions and the location of the arcs that fill some of them.

ONCE UPON A RING

The origin of planetary rings is much debated, and the fundamental question is whether they were formed at the same time as the planet, or later as the result of an intense collisional evolution of larger primordial bodies. The lifetime of ring systems is also debated – in particular, the long-term stability of their present structure, involving gaps, arcs, resonances and moonlets.

Some of the rings evolve rather quickly, and vanish within a few million years or even less. One of the driving mechanisms is the frequent collisions occurring within densely populated rings, which lead to a diffusion of the ring particles both inwards and outwards with respect to the planet.

There are attractive analogies with the protoplanetary disks observed around young stars, or the accretion disks sometimes characterising binary systems, but they do not account for the diversity among the observed ring systems. As an example, one of the most credited hypotheses on the formation of Saturn's rings foresees a past scenario in which a giant comet smashed into one of the larger moons and completely shattered it. The debris originated by such an event would then have been rearranged, mainly by celestial mechanics, into rings and moonlets.

In order to reconcile the different views, an overall distinction between primary and secondary rings has been introduced. To the former type belong the bright densely populated disks consisting of relatively large particles (up to a few metres), while secondary rings are darker and composed of micrometric dust particles. The dynamics of the primary ring structures is dominated by collisions and gravitation, while the small particles characterising secondary rings are also subject to non-gravitational forces such as those induced by the magnetic field of the planet or by solar radiation pressure.

Modelling all these different interactions in order to investigate the origin and evolution of planetary rings is still subject to large uncertainties, but it is of fundamental interest in planetary science. Within the rings, all the basic forces responsible for planetary formation in the early Solar System can be observed in action and on relatively short timescales.

10

At the edge of the Solar System

Imagination is more important than knowledge.

Albert Einstein

Understanding the nature of the transneptunian objects represents the last frontier of planetary science - and not simply because we are dealing with celestial bodies orbiting at the edge of the Solar System. New surprising discoveries are continuously posing stimulating questions for celestial mechanics: the dynamics of unusual binary systems and of resonant large-sized bodies, their role as parent bodies of most short-period comets. In the ten years since the discovery of the first TNO, the population has grown sufficiently to allow preliminary characterisation from a dynamical point of view. Yet the feeling is that we are only at the beginning of a long path that will eventually lead to bridging the gap between the last of the planets and the distant Oort Cloud.

BEYOND PLUTO

For a long time, Pluto was the only known celestial body orbiting at the border of the Solar System. Far in the distance, a thousand times farther, towards the closest star, the Oort Cloud of comets marks the limit of the gravitational domain of the Sun. But what is likely to be found in between? More planets, empty space, or a ring of natural debris? This was the recurring question after Clyde Tombaugh discovered Pluto in 1930.

The answer was left to the theoretical modelling of the early phases of planetary formation. In 1949 Kenneth Edgeworth, and in 1951 Gerard Kuiper, independently suggested the existence of a disk-shaped region, composed of small icy bodies, beyond the orbit of Neptune. In this respect they would represent valuable primitive bodies, as the planetary accretion process at the outer edge of the Solar System could have been stopped at an early stage, thus bearing some similarity to the asteroid belt between the orbits of Mars and Jupiter. This hypothesis was recently borne out by computer simulations showing that the 'belt' was consistent with the overall scenario of the origin of the Solar System, foreseeing the cooling of a protosolar nebula. Over the ensuing

How celestial bodies are named

The rules for numbering and naming the small bodies of the Solar System were established by the International Astronomical Union, and are implemented at the Minor Planet Center, in Cambridge, Massachusetts.

When a candidate new object is observed it is given a provisional designation, as follows. The first four digits are the year of discovery, after which capital letters and sometimes numbers appear. (Only 25 letters are used, as the letter I is excluded to avoid confusion with the number 1.) The first letter refers to the half-month of the observation: the letter A denotes January 1–15, and B denotes 16–31 January, and so forth). This is followed by another letter assigned sequentially, thus denoting the order of discoveries within the half-month. If a half-month is particularly rich in discoveries and all the letters are used, a final number is added to indicate the number of times the second letter has been used. As an example, 2005 GF187 was discovered in the year 2005 during the first half of April, and was the $(25 \times 187) + 6 = 4{,}681$th object observed during that period.

This might seem a rather complicated procedure, but it was first implemented in 1925, when discoveries were not as frequent as they are today. Furthermore, it happens that after checking previously discovered objects, not all candidates are new discoveries. However, if the candidate *is* a true discovery, a catalogue number is officially assigned by the MPC only after its orbital elements are determined with sufficient accuracy to allow safe recovery. After that, the discoverer(s) have ten years to propose a name, which must be approved by the IAU Committee on Small Bodies Nomenclature.

There are also additional ethic and programmatic rules. A planet should bear the name of a major deity (an Edgeworth–Kuiper Belt object is named after the gods of creation), while main belt asteroids have a wider variety of names: nations, towns, poets, musicians, scientists and astronomers, either historical or current. (Politicians should be left until 100 years after death, in order to allow history to provide the necessary judgement on their actions.) Apart from avoiding duplication or offensive terms, the names of pet animals are also officially discouraged.

Within this framework, Jewitt and Luu expressed the wish to name their discovery 'Smiley' (after the character in John Le Carré's spy novels) because it had been elusive for such a long time. Unfortunately, the name had been already used for asteroid number (1613), and 1992 QB1 is still unnamed, although it has the catalogue number 15760. It is nevertheless a remarkable coincidence that the suffix QB1 was assigned solely on the basis of the timing of its discovery, although it seems to be an abbreviation of 'Kuiper Belt n.1'.

years these results were confirmed by more refined modelling and by indirect dynamical clues, such as the excess of comets in short-period orbits (mentioned in Chapter 5).

Although on scientific grounds there appeared to be no reason for the planetary system to end at Pluto, the direct observation of faint bodies orbiting in the region identified by Edgeworth and Kuiper was out of reach for ground-based observations until the mid-1980s, when the introduction of CCDs provided the required technological breakthrough.

On 30 August 1992 the American astronomers David Jewitt and Jane Luu, using the 2.2-metre telescope of the University of Hawaii on Mauna Kea, discovered a celestial body on an orbit just beyond that of Pluto, at a mean distance from the Sun slightly in excess of 40 AU. The object was provisionally designated 1992 QB1. It is very dark, with a reddish colour indicating the presence on its surface of organic compounds – typical of icy bodies (such as comets) that have long been exposed to cosmic rays. Object 1992 QB1 was therefore the perfect candidate for being the first member of the long-sought population of Edgeworth–Kuiper Belt objects (EKBO).

The discovery of 1992 QB1 was the first of a long series, and new celestial objects began to be commonly referred to within the astronomical community as *cubewanos*. We now know that beyond Pluto the Sun is surrounded by a large ring of frozen icy worlds, mostly concentrated in the region between 39 and 56 AU. They are very dark objects reflecting only a few percent of incident sunlight, and seem to share with comet nuclei other peculiar physical properties. However, they are larger than comets, with diameters ranging from hundreds to some thousands of kilometres, thus being comparable in size to Pluto and Charon. Most of them travel on almost circular orbits with small inclinations to the ecliptic plane, but there are also high-eccentricity objects reaching to distances as far as 1,000 AU from the Sun. It is estimated that the Edgeworth–Kuiper Belt contains more than 100,000 objects with diameters larger than 100 km. If this estimate is confirmed, the total mass of the bodies contained in this region is much larger than that presently inside the asteroid belt between Mars and Jupiter.

As the number of known EKBOs grew, becoming a statistically significant sample of the whole population, it could be clearly seen that it had a peculiar structure which could not be brought back to a simple 'belt' (Figure 10.1). As a consequence it is now common practice to indicate, with the generic term 'transneptunian objects' (TNOs), every celestial body orbiting beyond Neptune, leaving the term EKBO referring only to bodies with density peaking around 40 AU.

EKBOs also constitute a source for the population of short-period comets, especially those with a period of revolution of the order of 10 years (dynamically controlled by Jupiter). If this is true, it is expected that some of these objects will be found well inside the planetary region, thus witnessing the slow diffusion process from the Edgeworth–Kuiper Belt to the inner Solar System. Indeed, a group of relatively large objects – the Centaurs –presently exists on highly eccentric unstable orbits between Jupiter and Neptune.

Many scientific investigations have been devoted to the study of the formation of the EKB; for example, H.F. Levison (Boulder, Colorado) and A. Morbidelli (Observatoire de la Côte d'Azur) conjectured that some EKBO formed between the orbits of Neptune and Pluto. Once again, mean motion resonances

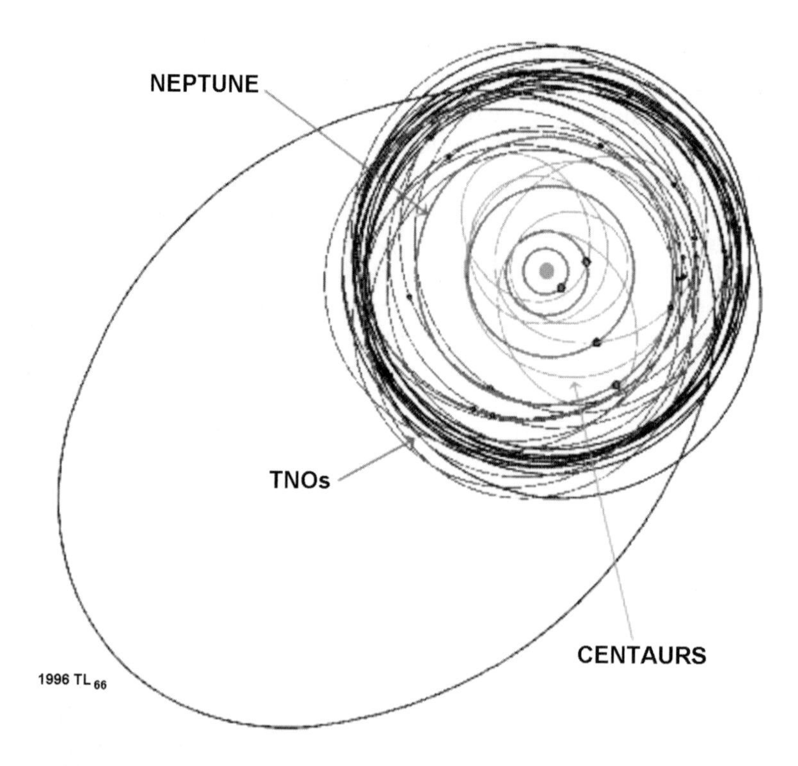

FIGURE 10.1. The distribution of the orbits of transneptunian objects. The EKB is located just beyond the orbit of Neptune. Peculiar objects such as the Centaurs and 1996 TL66 – a member of the scattered objects population – are also shown. (Courtesy Minor Planets Center.)

could have heavily shaped the Solar System by pushing these objects out to their present location.

SMILEY AND THE OTHERS

The boundaries of the Edgeworth–Kuiper Belt are not clearly defined, due to the presence of objects at distances as far as 500 AU from the Sun, and the limiting magnitude affecting telescopic surveys. Although observational biases could influence our present description of the belt, an overall picture begins to emerge.

Figures 10.2 and 10.3 show the distribution of the orbital eccentricities, e, as a function of the mean distances, a, from the Sun. It is easy to see that most bodies are concentrated in the region between 39 and 48 AU, with a peak at 39 AU, corresponding to Pluto's semimajor axis (39.48 AU). An internal structure of the belt is also recognisable, leading to the division of EKBOs into three major groups according to their dynamical features: *resonant* bodies concentrate around 39 AU, *classical* EKBOs are located between 42 and 48 AU, and *scattered* objects extend from 50 AU outward. On the other side, toward the inner Solar System, the high-eccentricity and high-inclination orbits of the Centaurs group around helio-centric distances of about 20 AU.

1992 QB1 - with its distance from the Sun varying between 40.9 AU at perihelion and 46.6 at aphelion - belongs to the densely populated region between 42 and 48 AU, characterised by low eccentricities and moderate inclinations. These objects probably formed there, thus retaining their original dynamics, and possibly perturbed by peculiar events occurring during the early phases of Solar System formation - perhaps the arrival of a large planetesimal ejected by the nearby planet Neptune, or a star passing in the neighbourhood of the Solar System.

The classical belt shows an abrupt edge at about 50 AU. Among the possible explanations, a star approaching very close, at an approximate distance of 150 AU from the Sun, could have contributed to a dramatic increase in the eccentricities and inclinations of the more distant transneptunian objects, eventually resulting in truncating the Edgeworth–Kuiper Belt at about 50 AU.

When dealing with the dynamics of distant objects, it has to be considered that our planetary system travels through the Galaxy, and, on astronomical timescales, can encounter massive bodies such as stars and giant molecular clouds. Moreover, the Sun probably formed inside a densely populated star cluster that has now dispersed, so that particularly close star-passages are likely to have occurred early in the history of the Solar System. The signature left by these strong gravitational perturbations are recorded in the high inclination and eccentricities of the perturbed objects.

This is one of the favourite explanations for the presence of the scattered objects with large eccentricities (Figure 10.2). The first discovered objects of this kind were named 1996 TL66 and 1996 GQ21, and they also exhibit large inclinations of the order of 20°. More recently, the TNO 2004 XR190, named Sedna, has puzzled astronomers because of its extremely distant aphelion reaching to 1,000 AU. An extreme hypothesis has been formulated: could it be an object belonging to another star, left behind during a close encounter with our Sun? If this is the case, then Sedna can be considered as the closest extrasolar planet ever observed.

Orbital resonances are also present in the EKBO population. A consistent number of bodies located at 39 AU are apparent in Figures 10.2 and 10.3, as they are arranged in a vertical straight line. Sharing with Pluto, to a high degree of accuracy, the same mean distance from the Sun (see Table 10.1), they are also involved in 3:2 mean motion resonance with Neptune (discussed in Chapter 3). As a consequence, these resonant objects - called Plutinos - complete two revolutions around the Sun while Neptune completes three. As for Pluto, the resonance has a stabilising effect: the minimum relative distance with Neptune is always kept larger than 17 AU, thus preventing the Plutinos from approaching the planet dangerously close. Encounters between Plutinos are in general not effective, because of their small mass and dispersed inclinations (see Table 10.1),

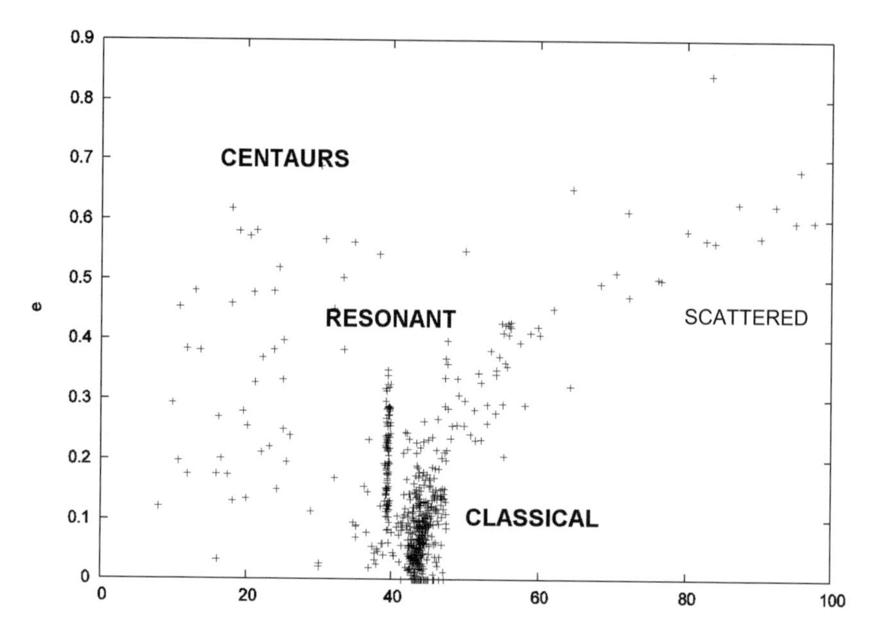

FIGURE 10.2. The distribution of eccentricities as a function of the heliocentric mean distance distinguishes the three different groups of EKBOs and the population of the Centaurs.

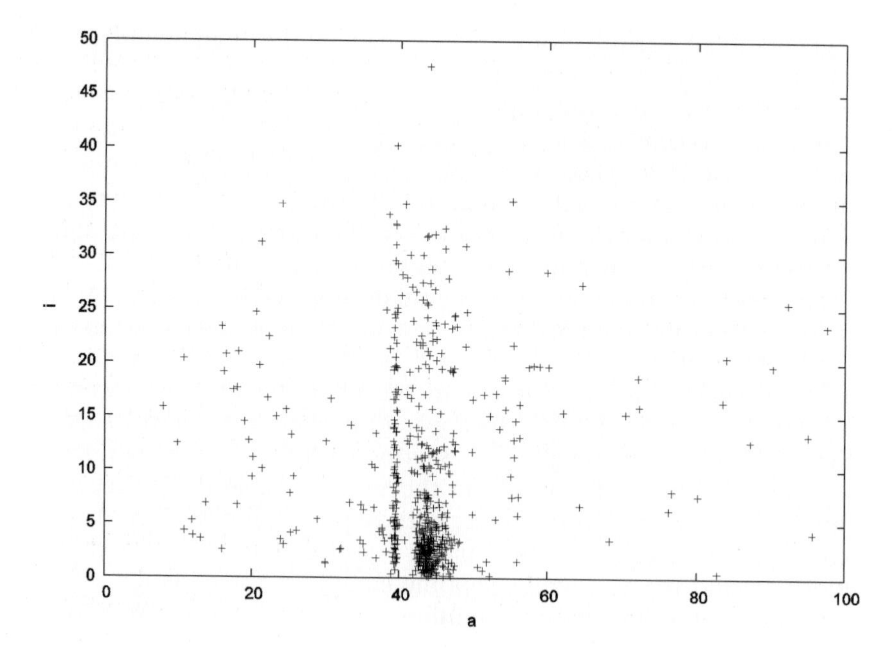

FIGURE 10.3. As in FIGURE 10.2, the same classification (resonant, classical and scattered) applies in the plot of inclination against heliocentric distances.

Table 10.1. Orbital elements of Pluto and some of the Plutinos.

	Semimajor axis (AU)	Eccentricity	Inclination (degrees)
Pluto	39.48	0.25	17.1
1993 RP	39.33	0.11	2.6
1995 HM5	39.85	0.26	4.8
1997 QJ4	39.20	0.22	16.6
1999 LB37	39.50	0.19	14.9
2001 QF331	39.48	0.11	2.7
2003 QV91	39.49	0.35	32.9
2005 GF187	39.50	0.21	3.9

which result in widening the separation between their orbital planes, so that strong perturbations are extremely unlikely.

BIG BROTHERS ARE WATCHING US

Another puzzling characteristic of the transneptunian objects population is the increasingly high number of large EKBO discovered to date. The first 'oversize' object was found in late 2000. Named Varuna, after one of the oldest Hindu gods, and catalogued as number 20,000, it has a classical EKB orbit and a diameter of about 900 km. In June 2002 the discovery of an object with dimensions about half that of Pluto (and therefore comparable to its satellite Charon) was announced: 2002 LM60 was found at about 43.5 AU on an almost circular orbit. This new EKBO was named Quaoar, after the great force of creation in the mythology of the Tongva people – a community of Native Americans from the Los Angeles area. It is now catalogued as number 50,000, and careful measurement of its dimensions by the Hubble Space Telescope revealed that it has a diameter of 1,250 km. There is also the aforementioned Sedna, with an estimated diameter of between 1,000 and 1,500 km.

The gap separating the largest EKBO from Pluto was shrinking fast; and it eventually disappeared when on 29 July 2005 an object possibly bigger than Pluto was identified: 2003 UB313 (now officially named 136199 Eris) has an orbit at almost 70 AU, with an eccentricity of 0.44 and an inclination of $44^{\circ}.2$.

Therefore, does Pluto deserve the status of 'planet', or should it be considered a large EKBO? The concerns raised by its chaotic dynamics (see Chapter 5) and by the discovery of the Plutinos had already led to the proposal of cataloguing Pluto among the small bodies at number 10,000; but this was eventually rejected on the basis of its much larger size (see Figure 10.4). However, with the discovery of 136199 Eris the question has again become very important, and the definition of 'planet' must be reconsidered.

Additional evidence against Pluto is the presence of its satellite Charon, with a

FIGURE 10.4. Relative sizes of the largest EKBOs compared to Pluto, and to some other well-known Solar System bodies. The object 2003 UB313 has been named 136199 Eris.

diameter half of Pluto's diameter. This is not an unusual situation, as many EKBOs, including 136199 Eris, have satellites. Moreover, Sedna even seems to be a contact binary: the two components are of comparable size and sufficiently close to eclipse each other during a full revolution.

The commonly accepted explanation for the frequent occurrence of binary systems among the EKBO is that collisions represented (and possibly still represent) a major evolutionary process in shaping the dynamical history of the entire population. A catastrophic collision is an extreme case resulting in the debris reaccreting into bodies of comparable size.

An active collisional scenario in the Edgeworth–Kuiper Belt is supported by an apparently distant sequence of discoveries. Since the end of 1997, tens of new satellites orbiting Uranus have been detected. Following the anomalous tradition adopted for the other satellites of the planet, they have been named after Shakespeare's characters such as Caliban, Sycorax and Prospero. As opposed to the large satellites of Uranus, they are tiny celestial bodies on highly irregular orbits. A possible explanation is that these new satellites are fragments of an EKBO captured by Uranus while on their way toward the inner Solar System.

Yet, against all odds Pluto seemed to be reaffirming its rights. In October 2005 – less than three months after having lost its supremacy in size over the EKBOs – two new satellites of its own have been discovered. And a satellite system is, after all, a planetary characteristic...

CHIRON AND THE CENTAURS

A number of objects named Centaurs – after the mythological beings, half human and half horse – are known to wander between the orbits of Jupiter and Neptune on elongated trajectories crossing the orbits of the outer planets. Their distribution is shown in Figures 10.2 and 10.3, while some orbital parameters are

Nemesis

The possibilities of what is to be found in the outer reaches of the Solar System has always intrigued astronomers. In 1984 – long before the discovery of the transneptunian objects – an article published in *Proceedings of the National Academy of Sciences* created a lively debate in the scientific community. Analysing paleontological data, David Raup and John Sepkosky claimed to have found a 26 million-year periodicity in the episodes of mass extinctions on our planet. It has been repeatedly pointed out that 'periodicity' is a keyword for celestial mechanics, and as a consequence an outburst of papers on the possible dynamical explanations appeared in the scientific literature. Some invoked the existence of an unseen massive solar companion orbiting in the extreme regions of the Solar System, as far as 88,000 AU, on a moderately eccentric orbit and with a period of revolution of 26 million years. At every perihelion passage this brown dwarf plunges inside the Oort Cloud of comets, perturbing their motion to the point of triggering intense cometary showers toward the inner Solar System. This dynamical mechanism would start large cometary impacts on our planet, with the consequent abrupt climate changes and the onset of mass extinctions. Because of the terrible influence on the history of our planet and the danger to the existence of the human species, this frightening object has been given the name Nemesis.

As often happens in science, the entire subject was soon abandoned, because the original data supporting the 26-million-year periodicity were placed into question. Whatever the truth, it is reassuring that celestial mechanics was able to compute that the next Nemesis-induced killer shower is foreseen a 'safe' 15 million years from now.

shown in Table 10.2. The Centaurs move on unstable orbits and suffer the gravitational influence of the giant planets, experiencing repeated close approaches and finally being ejected from the Solar System or transferred on orbits of much smaller size within a few million years. Due to their large orbital eccentricities they cross the orbits of the outer planets, as deduced by comparing the values of their perihelia and aphelia with the planetary distances (Table 10.2).

Centaurs are relevant for the dynamics of the Edgeworth–Kuiper Belt because they represent the evolutionary 'missing link' between EKBOs and short-period comets (Figure 10.5). Dynamical proof was preceded by physical observations which have puzzled astronomers for some time.

Chiron – the first Centaur – was discovered in 1977 by Charles Kowal (California Institute of Technology), and was classified as asteroid number 2060. It has peculiar orbital parameters, and during its perihelion passage in 1988 it displayed cometary activity by producing a coma and a tail. However, it is not a typical comet, because it is about 200 km in diameter – about fifty time larger than the nucleus of a short-period comet.

The discovery of bodies sharing similar orbital paths, and the study of the

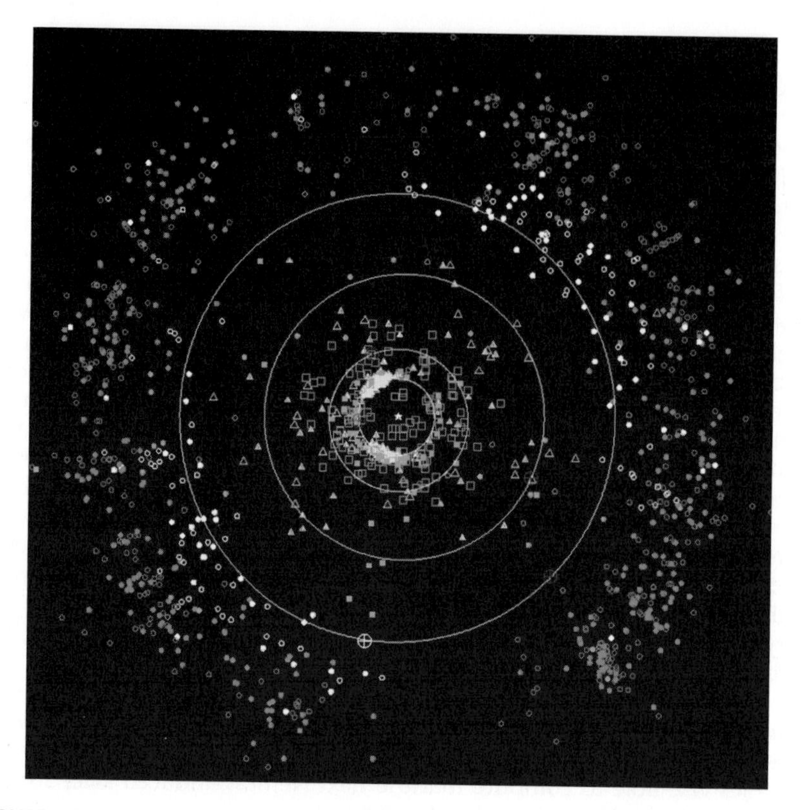

FIGURE 10.5. The diagram shows the orbits of the outer planets (from Jupiter to Neptune) and the position of the small bodies in the outer Solar System. The Edgeworth–Kuiper Belt lies beyond the orbit of Neptune. It is not dynamically isolated, and many transition objects are recognisable. The two densely populated groups located along the orbit of Jupiter (the innermost in the diagram) mark the position of the Greek and Trojan camps of asteroids. (Courtesy Minor Planet Center.)

long-term evolution of their orbits, has allowed us to place Chiron and the Centaurs in the proper framework of the diffusion of matter from the Edgeworth–Kuiper Belt to the inner regions of the Solar System. In a way similar to what has already been described for the asteroid belt, secular resonances are responsible for delivering EKBOs in the outer planetery regions, where their subsequent dynamical evolution is controlled by close encounters with the massive giant planets. As mentioned in Chapter 5, during these encounters the tidal stress undergone by a loosely bound icy nucleus is sufficient to cause the break-up of the body into several components, each the size of a typical short-period comet. Due to randomisation of the orbits caused by their chaotic nature, the different fragments belonging to the same parent body become dynamically unrecognisable.

Table 10.2. Orbital parameters of some Centaurs.

	Aphelion distance (AU)	Perihelion distance (AU)	Eccentricity
2060 Chiron	18.9	8.5	0.38
5145 Pholus	32.0	8.8	0.57
7066 Nessus	37.4	11.8	0.52
8405 Absolus	29.1	6.8	0.62
10199 Chariklo	18.7	13.0	0.18
31824 Elatus	16.3	7.3	0.38
32532 Thereus	12.7	8.5	0.20
49036 Pelion	22.7	17.2	0.14

PLANET X

At the beginning of the twentieth century the American astronomer Percival Lowell tried to find an explanation for the anomalies of the orbital motion of Uranus which could not be attributed to the gravitational perturbations by Neptune. His aim was to compute the orbit of an unknown Planet X, much in the same way as had been done for discovering Neptune (see Chapter 1). The letter X denoted 'unknown' rather than 10, as at that time no-one knew of the existence of Pluto. Although Lowell never succeeded in predicting a reliable position of the perturber, the quest for Planet X led to extended sky surveys, eventually resulting in the discovery of Pluto in 1930.

The ninth planet of the Solar System was obviously attributed with 'planetary mass', to account for the perturbations in the motions of Uranus and Neptune. As the years went by and more refined observations were carried out, Pluto's calculated mass and size decreased steadily, becoming lower than that of a terrestrial planet, until reaching the present estimate which places Pluto behind many natural planetary satellites, including the Moon. The search for Planet X was resumed once again, but not even the finding of the extended transneptunian belt could justify the Uranus–Neptune irregularities. Was there still an unknown planet awaiting discovery?

A sensible resolution was provided by the Voyager 2 measurements of the masses of the outer planets. Adopting the correct values and taking into account the unavoidable numerical errors introduced by modelling the motion of the planets, it is possible to justify the anomalies in the motions of Uranus and Neptune without invoking the influence of an additional body.

The quest for Planet X seems now to be over. Beyond Neptune – the outermost planet of the Solar System – a host of planetoids are busily orbiting with a crowd of large comets.

Planets and dwarf planets: the new Solar System

We have repeatedly pointed out throughout the text that while the transneptunian objects population grew and became more and more characterised both from a dynamical and a physical point of view, the commonly accepted definition of 'planet' encountered trouble. Shape, dimension and orbital motion no longer sufficed for granting planetary status to a newly discovered celestial body. In particular, it was realised that if Pluto were to be a planet, then the number of 'planets' in the Solar System would rapidly increase in the coming years. 136199 Eris is likely to be only the first of an unknown series of distant large TNOs whose presence would be soon revealed by advanced observational techniques. On the other hand it appeared a much too severe restriction to simply scale down Pluto and the largest TNOs to 'minor bodies' such as asteroids and comets (see Figure 10.4).

Intensive brainstorming within the astronomical community led to the approval – during the International Astronomical Union (IAU) general assembly held in Prague in summer 2006 – of an historical resolution which states that planets and other bodies in our Solar System be defined into three distinct categories:

- A 'planet' is a celestial body that (a) is in orbit around the Sun, (b) has sufficient mass for its self-gravity to overcome rigid body forces so that it assumes a hydrostatic equilibrium (nearly round) shape, and (c) has cleared the neighbourhood around its orbit.
- A 'dwarf planet' is a celestial body that (a) is in orbit around the Sun, (b) has sufficient mass for its self-gravity to overcome rigid body forces so that it assumes a hydrostatic equilibrium (nearly round) shape, (c) has not cleared the neighbourhood around its orbit, and (d) is not a satellite.
- All other objects, except satellites, orbiting the Sun shall be referred to collectively as 'Small Solar System Bodies'.

Reaching a final decision was not easy, and there was heated debate during the assembly. The key point which led to a large consensus was the consideration not only the present characteristics of a celestial body, but its whole 'life' from birth onwards. This is the basic meaning of the ability of 'clearing the neighbourhood'. A planet is a planet only if it has been successful in emptying it immediate region of space by ejecting or accreting onto itself the other objects (planetesimals, protoplanets, and so on) wandering around during the early stages of Solar System formation (as described in Chapter 5).

When it was not so, as observed in the asteroid belt and beyond Neptune, a certain number of larger objects remains dynamically embedded inside a population of smaller celestial bodies. A planet has not fully grown, thus justifying the new definition of 'dwarf planets'. To this new class belongs the round-shaped 1,000-km-diameter asteroid Ceres – the largest TNO known to date (136199 Eris) – and Pluto. A number of additional candidates exist, and there will be many in the future. A dedicated commission will decide on a case-by-case basis.

In order to avoid confusion, the IAU resolution also replaces the loosely-defined term 'minor bodies' of the Solar System with 'small bodies', now more fitting to reality.

As expected, Pluto's new status attracted a vast number of comments in favour or against the IAU resolution. It is left to the reader to decide whether Pluto should be historically considered the last of the planets or the very first member of the TNOs – a population which is significantly contributing to a deeper understanding of the origin and evolution of the planetary system.

11

On the road to exolife

> There is no fundamental difference between a living organism and lifeless matter. The complex combination of manifestations and properties so characteristic of life must have arisen in the process of the evolution of matter.
>
> *Alexandr Oparin*

The great expectations concerning the existence of evolved life-forms in the Solar System have not been confirmed by the images sent back by interplanetary probes. But we now know where to search for present or past signs: below the surface of Mars and the icy crust of Europa, among the frozen methane landscapes of Titan, and the active volcanic areas of Io and Triton. Moreover, at the end of the twentieth century a major discovery extended the traditional domain of planetary science, posing new problems for celestial mechanics and opening novel perspectives for the quest for life in the Universe. The detection of planetary systems around several stars has taken astronomers by surprise because of the unexpected variety of their dynamical configurations. And the next generation of space telescopes will probably answer the question of the existence of Earth-like planets in our Galaxy.

BEYOND THE SOLAR SYSTEM

Are we alone in the Universe? This is a crucial question that has always puzzled mankind; and in trying to answer it we are faced with many as yet unsolved problems about the nature of the Universe, the processes leading to the birth of stars and planets, and the intimate structure of our Solar System and the planet on which we live.

The development of terrestrial-like life-forms needs a suitable environment: a relatively small rocky planet equipped with water, an atmosphere, and all the basic life-sustaining biochemistry (such as methane and amino acids). Physical and orbital characteristics must remain stable, and this mostly depends on the type of star illuminating the planet, on the planet's distance from its 'sun', and on the size and arrangement of other celestial bodies within the system (for example, resonances and/or Bode-like laws).

We know that a non-zero obliquity of a planet leads to seasonal cycles that have positive side-effects, but we are also aware that beyond certain limiting values it can lead dangerously to chaotic spin–orbit regimes resulting in dramatic climatic changes on a global scale. And what about the presence of a large moon, whose many influences on the dynamical and biological evolution of the Earth have been repeatedly emphasised?

At first glance the correct combination of so many ingredients, involving celestial mechanics as well as many other disciplines, might seem so peculiar that it is easy to believe that the Earth is an extraordinary and unique planet for hosting life. Yet the number of stars in the Galaxy and the number of galaxies in the Universe is so high that it is a difficult concept for the human mind, and for a long time this has prevented any reliable estimate of the chances of finding Earth-like planets around other stars.

The existence of worlds similar to the Earth was conjectured by Epicurus around 300 BC, and his vision was shared by Titus Lucretius Carus (first century BC), who wrote the following in his *De Rerum Natura* (translated by Cyril Bailey, Oxford, 1947/1986):

> That this one world and sky was brought to birth,
> But that beyond it all those bodies of matter do nought;
> Above all, since this world was so made by nature,
> And the seeds of things themselves of their own accord,
> Jostling from time to time by chance,
> Were driven together in many ways,
> Rashly, idly, and in vain,
> At last those united, which, suddenly cast together,
> Might become ever and anon the beginnings of great things,
> Of earth and sea and sky,
> And the race of living creatures.

Conversely, Aristotle was firmly convinced of the uniqueness of the Earth, and placed it at the very centre of the Universe – an idea that strongly influenced science and philosophy for many centuries. The geocentric–anthropic view became religious dogma. In the sixteenth century Giordano Bruno was burned at the stake, on the basis of his belief of the existence of innumerable worlds in the Universe. Modern scientists tried to look for experimental evidence of the existence of planetary systems around nearby stars. In the seventeenth century the Dutch scientist Christiaan Huygens attempted direct observations, but was unsuccessful because his telescopes were unable to separate the light of a possible planet from that of the star – a technological limitation that still exists today. The British astronomer Arthur S. Eddington was deeply convinced of the existence of extrasolar planetary systems, and in 1933 he went as far as declaring that it was an absurd belief that Nature is not busily repeating, somewhere else, the same strange experiment successfully performed on Earth.

Earths in the sky

The book *Les Terres du Ciel* (*Earths in the Sky*) – written by the French astronomer and educator Camille Flammarion (1842–1925), and first published in 1884 – reflects the hopes of the late nineteenth century for the existence of extraterrestrial life. Contemporary knowledge of the time, about the planets of the Solar System, is extensively illustrated with beautiful drawings and photographs. At the end of each chapter devoted to an individual planet, the important topic of possible inhabitants is treated. But it is not mere fantasy, because the development of alien habitats is extrapolated on the basis of real astronomical data. As an example, the proposed landscapes on Venus have a typical 'Mediterranean' style, and Flammarion remarks that the absence of a large moon, coupled with insufficient solar tides, would not lead to the periodic raising and lowering of the oceans – a 'scientifically correct' hypothesis ruled out by the Russian Venera missions, which measured the ground temperature: around 500 °C.

In Flammarion's vision, Mars is green with every species of tree and plant, watered by the network of channels covering the entire surface of the planet (Figure 11.1). In the evening twilight the first bright star appearing over the horizon is our planet, the Earth (Figure 11.2).

Any attempt to determine what kind of life could develop under the immense atmosphere of Jupiter is much more difficult, but Flammarion's description of the Galileian satellites passing in front of large openings in the jovian clouds closely matches the spectacular images returned by the Voyager spacecraft in 1979.

It is worthwhile noting that the chapters in Flammarion's book do not follow the arrangement typical of astronomical textbooks in which the planets are ordered by increasing distance from the Sun, but rather on their affinity with the Earth; that is, their habitability – a concept now in the front line of astrobiology.

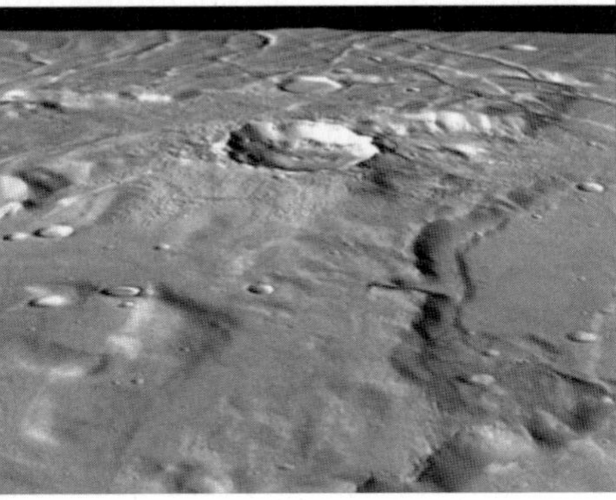

FIGURE 11.1. Flammarion's vision of martian channels, and a perspective view of the martian Solis Planum region as imaged by Mars Express. (Copyright ESA/DLR/FU Berlin (G. Neukum).)

FIGURE 11.2. The Earth sparkles in the martian evening sky, as imagined by Flammarion (left) and as imaged by the NASA Exploration Rover Opportunity in 2004. (Courtesy NASA/JPL.)

HUNTING FOR EXOPLANETS

On 6 October 1995 the Swiss astronomers Michel Mayor and Didier Queloz announced the discovery of the first extrasolar planet (exoplanet) – a celestial body about half the mass of Jupiter, orbiting the star 51 Pegasi. It was the end of a longstanding quest, and since then a plethora of exoplanetary systems have been found. To date more than 150 have been discovered, and this number will certainly grow quickly in the coming years, especially after the placing in orbit of space telescopes such as those developed for the European Gaia and the French Corot missions.

The reasons why the expectations of Epicurus, Lucretius, Giordano Bruno, Sir Arthur Eddington and many others were only recently fulfilled are essentially twofold: the different nature of a star with respect to a planet, and the observational challenge posed for astronomers.

A star is an immense nuclear-powered engine emitting light by its own means. A celestial body becomes a star only if its mass is large enough to reach the extremely high pressure and temperature needed to trigger and sustain nuclear fusion reactions in its interior. Planets are celestial bodies which do not possess such a large mass, and they would appear only as dim and dark worlds were they not illuminated by the light coming from a nearby star. The difference between a large celestial body emitting light all over its surface and a much smaller body reflecting only a tiny fraction of that very same light translates into a wide gap in their relative brightness – a factor of around 1 billion.

Then comes the distance. The star closest to the Sun is α Centauri, about 4 light-years away – more than 250,000 AU. An hypothetical planet around α Centauri would be more than 8,300 times farther than the distance between the Earth and Neptune.

From an observational point of view these considerations imply the ability to distinguish two celestial objects having an angular separation extremely small and characterised by extreme contrast in their relative magnitude. This places severe requirements on optical telescopes for direct imaging. Exoplanets are hidden from our eyes by the overwhelming light of their suns. How, then, were extrasolar planetary systems unveiled in 1995?

Indirect methods of detection have proven very effective. Instead of trying to 'see' the planet, its presence is revealed by measuring the anomalies in the motion of the host star. It is well known that the stars in our Galaxy are not still, but move according to gravitation, thus very slowly changing their relative positions in the sky. Even though it will take 100,000 years for the familiar shapes of the constellations to be significantly altered (Figure 11.3), the proper motion of an individual star can be traced over the years by careful astrometric observations (precise positional measurements). The presence of a planet produces a characteristic periodic signature in the trajectory of the star, as its attraction slightly displaces the star's position as the planet completes a full revolution (Figure 11.4). Obeying Newton's law, the amplitude of this variation strongly depends on the mass of the planet and on its mean distance from the host star. The bigger and closer the planet, the larger the deviations.

As an example, the gravitational force exerted by Jupiter on the Sun is about 11 times stronger than that of Saturn. On the other hand, the pull due to the Earth and to Saturn are comparable in magnitude because Saturn is much bigger than the Earth but our planet is closer to the Sun.

Another widely used technique is the radial velocity method, which focuses on the change of velocity of a star measured along the observer's line of sight. The appearance in the data of characteristic periodicities can again be taken as an indication of the presence of a planetary system.

In spite of the relatively simple theory behind this method it is not easy to produce reliable results from astrometric observations. The variations of the proper motions of the stars are extremely small, and in order to detect them the

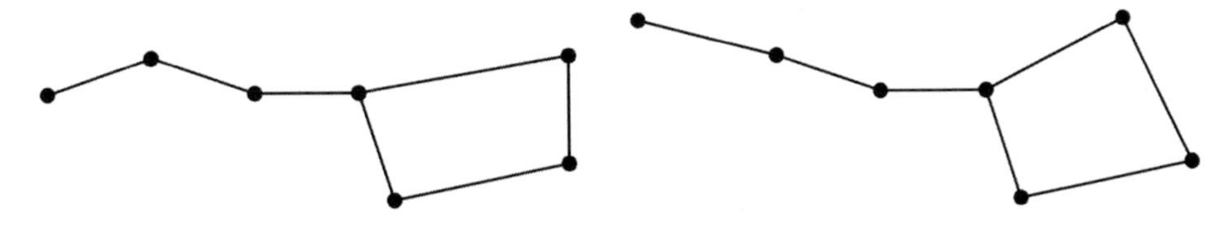

FIGURE 11.3. The Plough as it appears today (left), and its shape 50,000 years ago (right).

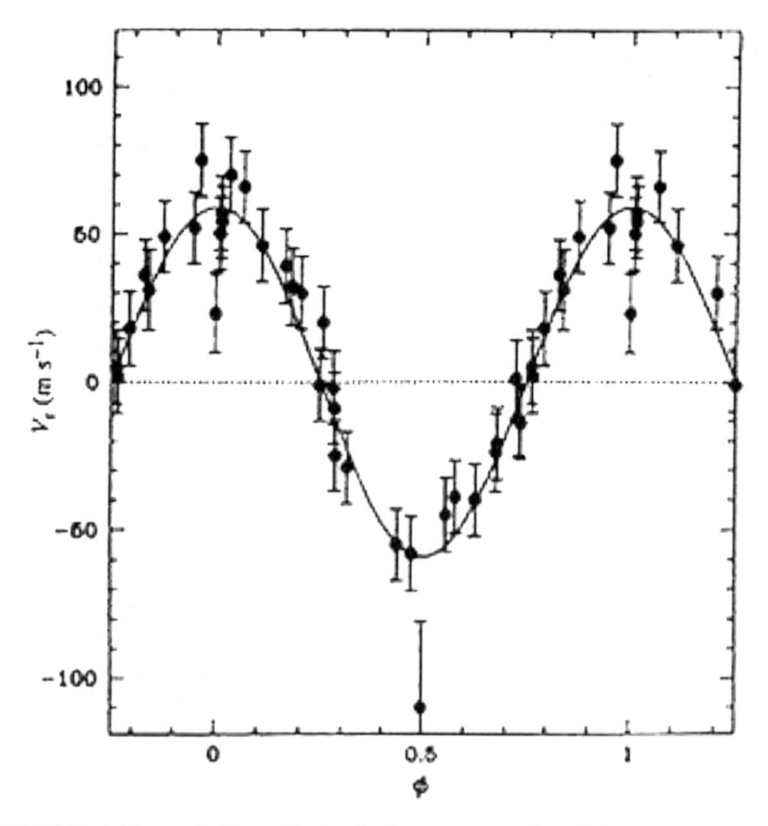

FIGURE 11.4. The periodic oscillation in the proper motion of the star 51 Pegasi revealed the presence of the first extrasolar planet to be discovered.

data must cover a significant fraction of the orbital period of the planet. It is not surprising, then, that the majority of discoveries are of large planets with short revolution periods, orbiting very close to their host stars (Table 11.1).

Other indirect methods have therefore been developed. One of the easiest to apply – especially in view of the possibility of observing from space – is based on a well-known astronomical event which occurs in our Solar System: the transit of a planet in front of the Sun. As seen from the Earth the dark disk-shaped profile of the planet appears like a circular shadow moving across the surface of the Sun (Figure 11.5). In this respect a transit can be considered as a scaled-down eclipse, because a small fraction of the incoming sunlight does not reach the Earth. In our planetary system these events involve only Venus and Mercury, because their orbits lie within Earth's orbit, thus allowing their passage between the Earth and the Sun. But the geometry of a transit can in principle be achieved by observing another star when one of its planets crosses the line of sight. In this case, even if it is not possible to 'see' the planet, the slight decrease in the luminosity of the star can be measured, and by also measuring the duration of the event we can produce an estimate of the size and orbit of the candidate exoplanet.

FIGURE 11.5. A transit of Venus in front of the Sun.

Pulsating planets

Historically the first detection of an exoplanet dates from 1992, but the celestial bodies involved were very far from the common perception of a star and a planet. The technique was to analyse the signals emitted by some of the most exotic members of our galaxy: *pulsars*. This name is a contraction of 'pulsating star', which was originally used by astronomers to identify the sources of strange radio emissions in the sky, which were repeating at regular intervals of time. Once their origin as 'messages' from alien civilisations was excluded, astrophysics found a convincing explanation for them as a by-product of stellar evolution. A pulsar is a rapidly rotating and very compact object – the surviving core of a massive star that has ended its life in a cosmic explosion: a spectacular *supernova*.

Within this framework, the existence of planets orbiting a pulsar can be deduced from the timing measurements of the radio signal which again may show distinct periodicities. After some false alarms the American astronomers Alex Wolszczan and Dale Frail eventually succeeded in producing the first convincing evidence of the existence of planets orbiting around the pulsar PSR1257+12, located some 1,000 light-years away in the constellation Virgo. Celestial mechanics played a deciding role, because confirmation of the discovery was made possible due to the existence of a 2:3 mean motion resonance of two of its planets, which produced a characteristic fingerprint in their mutual perturbations.

A GALACTIC ZOO

Within ten years of the discovery of the first exoplanet, about 150 planetary systems around other stars had been detected, mostly with astrometric techniques. Some physical and orbital data of a representative sample are shown in Table 11.1. Their mass M_J is measured in units of Jupiter's mass (a planet more massive than Jupiter has M_J larger than unity, while if it is smaller M_J is between zero and 1), and must be considered as a lower-limit estimate. Figure 11.6 shows statistics on the population of exoplanets known to date. Exoplanets take the name of the central star followed by a sequential label (b, c, d...).

Initially only large and isolated planets were found, but with the refinement of detection methods multiple planetary systems – such as those orbiting Gliese 876 and HD 12661 – could be identified.

Exoplanets are characterised by a wide variety of masses and dynamical behaviour. Bodies ranging in size from those slightly larger than Earth to giant planets larger than Jupiter have been found orbiting closer to their 'sun' than Mercury, as well as farther away than Neptune. The eccentricity of their orbits also exhibits large variations. Some are on almost circular orbits, while others

Table 11.1. Physical and orbital data of some exoplanets.

	M_J	Period (days)	Semimajor axis (AU)	Eccentricity
ε Eridani b	0.86	2,502.1	3.3	0.608
Gliese 876 b	1.935	60.94	0.21	0.025
Gliese 876 c	0.56	30.1	0.13	0.27
Gliese 876 d	0.023	1.93776	0.021	0.0
HD 142 b	1.0	337.112	0.98	0.38
HD 80606 b	3.41	111.78	0.439	0.927
HD 202206 b	17.4	255.87	0.83	0.435
OGLE-TR-56 b	1.45	1.21	0.022	0.0
55 Cnc b	0.784	14.67	0.115	0.0197
55 Cnc c	0.217	43.93	0.24	0.44
55 Cnc d	3.92	4,517.4	5.257	0.327
51 Pegasi b	0.468	4.231	0.052	0.0
OGLE-TR-10 b	0.54	3.10	0.042	0.0
TrES-1	0.61	3.03	0.039	0.135
HD 209458b	0.69	3.52	0.045	0.07
PSR B1620-26	2.5	100 yrs	23.0	–
2M1207 b	5.0	–	55.0	–
OGLE-05-071 b	2.7	2,920	3.0	–
HD 12661 b	2.3	263.6	0.83	0.35
HD 12661 c	1.57	1,444.5	2.56	0.2
HD 192263 b	0.72	24.348	0.15	0.0
HD 2638 b	0.48	3.444	0.044	0.0

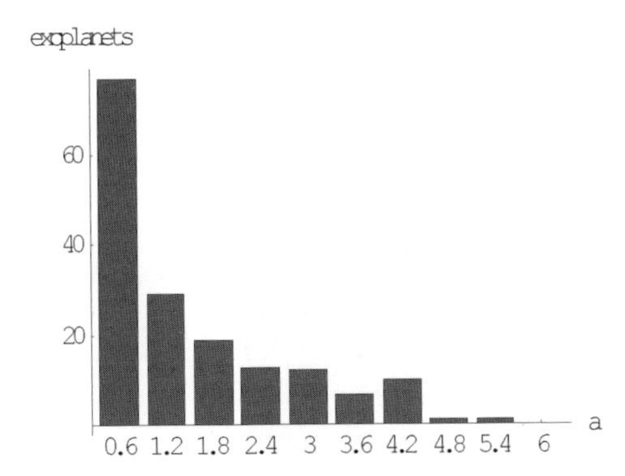

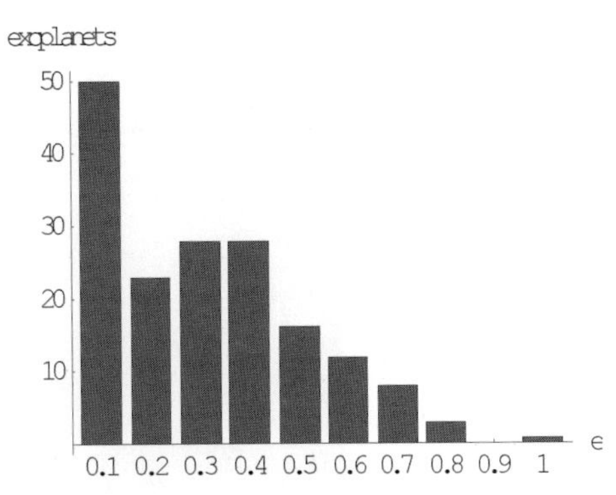

FIGURE 11.6. Histograms showing the distribution of semimajor axis (left) and eccentricity (right) for all the currently known exoplanets.

move on extremely elongated trajectories – such as HD 80606 b, the orbit of which approaches dangerously close to the parabolic limit ($e = 1$).

A puzzling consideration is the large fraction of Jupiter-sized planets orbiting extremely close to the star, as in the case of 51 Pegasi b, with a semimajor axis as low as 0.05 AU – a tenth of Mercury's distance from the Sun. This is astonishing, because according to the commonly accepted scenario of planetary formation, giant planets should appear only beyond a certain distance from the star in order to retain their immense gaseous atmospheres. As an example, in our Solar System there is a sharp border – the asteroid belt – separating rocky and gaseous planets. A possible explanation is that these 'hot Jupiters' formed at the 'correct' distance and then slowly migrated inward.

Another problem is that many exoplanets move on highly eccentric orbits. Figure 11.6 shows that only a third of them have an eccentricity lower than 0.1. This is again difficult to explain, as it represents a non-trivial extension to the general problem of the long-term stability of the Solar System (as discussed in Chapter 5).

Applying to the data in Table 11.1 the same procedure described in Chapter 3 for finding commensurable motions among Solar System bodies, it is possible to see that orbital resonances are frequent among exoplanets. Gliese 876 b has a period of 60.94 days – almost double that of the nearby planet Gliese 876 c (30.1 days). A similar approximate relationship holds within the systems surrounding HD 82943 and HD 128311; 55 Cnc b and 55 Cnc c provide an example of a 3:1 exoplanetary resonance; and HD 202206 exhibits a period ratio of about 5. Following Sylvio Ferraz-Mello (University of San Paulo), exoplanetary pairs can be classified according to their orbital period ratio, corresponding to a strong, moderate or weak interaction. In the first case the period ratio is small and resonances can occur, while in the latter case the period ratio is high and the orbits of the planets can be far from each other, presenting a greater chance of forming a stable system.

The next few decades will undoubtedly be marked by significant advances in

FIGURE 11.7. The brown dwarf 2M1207b and its planetary companion.

the investigation of extrasolar planetary systems with the development of novel detection techniques and the extablishment of dedicated infrastructures both on Earth and in space. Answers to the many problems on the structure of extrasolar planetary systems will also allow a deeper understanding of the formation and evolution of our own planetary system, while for celestial mechanics it represents a new and exciting scenario.

Imaging exoplanets

An example of the observational bias affecting the detection of exoplanets is the case of 2M1207b. Discovered by a group of European astronomers toward the end of 2004, it was a $5M_J$ planet orbiting its star at twice the distance of Neptune from the Sun. For the first time the detection was not obtained by indirect methods, but by resolving both celestial bodies with the 8.2-metre Very Large Telescope of the European Southern Observatory in Chile. Can 2M1207b be considered the first image of an exoplanet? The answer is 'yes' and 'no' at the same time. Firstly, the planet is orbiting around a brown dwarf – a very dim type of star that does not support nuclear fusion – a 'failed star'. However, its temperature is sufficiently high to make it visible mostly at infrared (thermal) radiation wavelengths, which are not visible to the human eye. The discovery was therefore made possible by obtaining an infrared image of the system (Figure 11.7). The planet is also sufficiently massive to be considered a failed star; and its temperature is fairly high, reaching up almost 2,000° C. Should we therefore consider the two celestial bodies as a star–planet pair or as a binary failed-star system? Whatever the answer, it is clear that direct observation of both celestial bodies has been possible only because of the low brightness of the primary 'star' and by the large distance between the pair.

Yet the most important result is by now well assessed – the observational evidence of what scientists and philosophers have considered since antiquity: planet formation is not an occasional event, but extends throughout our Galaxy and the entire Universe.

ASTROBIOLOGY

When discussing exoplanets a fundamental question becomes unavoidable. How many of the discovered bodies can in principle host life? The answer is at present mostly negative. Jupiter-sized planets orbiting as close as Mercury to their star have little in common with the Earth. Surprisingly, planetary systems orbiting pulsars more closely resemble terrestrial planets, if it were not for the fact that they are dark worlds gravitationally bound to a cold supernova remnant – possibly bodies that reaccreted after the explosion of the parent star. Yet the situation is not as bad as depicted by these considerations, since the limitations intrinsic to our detection techniques are in favour of finding massive planets orbiting at large distances from the host star. The discovery of Earth-like planets within planetary systems more similar to our own is still to come.

In general, for the existence of extraterrestrial life on a planetary scale a number of basic conditions must be met. A carbon-based organic chemistry requires a terrestrial planet orbiting inside a certain region of space termed the *habitable zone*. The problem is in defining the extent of such a region, and a possible criterion for tracing its inner and outer boundaries is to observe the intensity of the incoming radiation, which should be compatible with sustaining liquid water on the surface of the planet and an atmosphere around it. If applied to our Solar System, the habitable zone should not exceed 5 AU from the Sun, thus including Mars. One of the reasons why it is so important to find even the smallest trace of past life on the Red Planet is that it would confirm that the onset of biological evolution is not restrained to a narrow region centred at 1 AU from the Sun. If Mars were more massive, it could have possibly developed an active geology and retained an atmosphere, thus fulfilling the basic requirements for the development of life.

On the other hand, there are strong indications of the presence of oceans of liquid water under the frozen crust of some of the icy satellites of the outer planets such as Europa and, more recently, Enceladus. If this finding is confirmed and evolved life forms are discovered, then the habitable zone widens dramatically and life becomes a phenomenon potentially widespread across the Universe.

This is why the direct observation of exoplanets is essential in order to assess the presence of life. Being able to distinguish the light coming from the planet and that of the companion star implies the possibility of analysing them separately to search for characteristic signatures of life, such as the detection of molecular oxygen. In this context, astrobiology (or exobiology) is a recently-born discipline that merges astronomy, biology, chemistry and geology with the aim of accruing the necessary body of knowledge, allowing us to recognise the existence of life beyond our planet.

The shock of life

One of the first consistent scenarios for the appearance of life on our planet was proposed by the Russian biochemist Alexandr Ivanovich Oparin (1894–1980) in his treatise *The Origin of Life*, published in 1938. He conjectured that life appeared early in the history of the Earth, when the atmosphere was completely different from the present one, being rich in methane, ammonia and water. From these raw materials the combined action of heat and powerful electric discharges provided by lightning produced complex carbon compounds such as amino acids – the building blocks of living organisms.

Following Oparin's ideas, in 1953 Stanley Miller – a chemist at the University of California – performed laboratory experiments in which electric charges were passed through a mixture of methane, ammonia, water and hydrogen. The results were astonishing. In a few days more than 15% of the carbon was transformed into amino acids and other organic elements. However, Miller's experiment was criticised by the scientific community. The composition of the primordial Earth's atmosphere could only be tentatively guessed, and no evidence of organic chemistry more complex than amino acids was provided. In the second half of the twentieth century, several alternative theories were proposed, and recently the NASA Ames Research Center reported the formation of amino acids obtained by radiating icy bodies with ultraviolet light, thus reproducing a typical deep-space environment.

When focusing on intelligent life, the debate intensifies. From a philosophical point of view, the anthropic principle, in its strong formulation, states that the Universe must intrinsically allow the development of living species able to observe it. Unfortunately this can lead to opposite conclusions: that intelligent life is everywhere in the Universe, and conversely, that it is a unique phenomenon.

A more practical approach is that followed by the American astronomer Frank Drake in providing an estimate of the probability of finding extraterrestrial civilisations possessing the technology for communicating with each other. Drake wrote an equation which was presented at the first SETI (Search for Extra-Terrestrial Intelligence) conference held in 1961 at Green Bank, West Virginia. Indicating with N the number of civilisations which are expected to have developed interstellar communications, the Drake equation takes the form:

$$N = N_g \times N_s \times f_P \times f_L \times f_I \times f_T \times f_S$$

The key quantities are as follows:

- N_g is the number of galaxies in the Universe.
- N_s is the number of stars in a Galaxy.
- f_P is the fraction of stars with habitable planets.
- f_L is the fraction of planets where life actually develops.
- f_I is the fraction of planets on which intelligent life evolved.
- f_T is the fraction of life sites developing communications technology.
- f_S is the fraction of technological civilisations presently surviving.

Some of the individual terms involved have reasonably known values: for example, the average number of stars in a galaxy is estimated at 100 billion. But other quantities are subject to consistent variations as our knowledge improves, or will probably be estimated in the near future. Advanced observational techniques will presumably soon lead to the discovery of a statistically significant sample of exoplanets within the habitable zone.

The last three terms in Drake's equation are the most difficult to evaluate, especially because they are somehow tightly linked to each other. Intelligent life-forms are characterised by their ability to communicate, and the exploitation of radio signals is a long-range extension of this basic skill. Unfortunately, radio signals propagate with the speed of light (300,000 km/s), so that 'traditional' interactive communication is prevented by the immense distances between the stars (from tens to thousands of light-years and more). Unless we achieve a breakthrough in our understanding of the fundamental physical laws governing the Universe, foreseeing the possibility of sending signals faster than light, our best hope is to detect intelligent life without the chance of any high-level two-way communication. In particular, f_S is a measure of the lifetime of communicating civilisations - a concept difficult to grasp, because one is forced to take into consideration the possibility that technology will not, in the long term, be able to prevent mankind from following the fate of other species on our planet. If much too high a value is given to f_S, the obvious question is why the Earth has not yet been visited by an alien civilisation possessing a technology a million years ahead of ours. And if this value is too low, communication between intelligent life forms is forever prevented, because none of them survives long enough to develop the required interstellar communications technology.

In spite of the many unknowns, the Drake equation is a successful attempt to highlight the key topics that must be considered in trying to answer the question with which we began. Are we alone in the Universe?

BACK TO THE FUTURE

Whatever the answer to the ultimate question of life in the Universe, the future could soon bring the first direct observation of Earth-like extrasolar planets. In order to overcome the 'blinding effect' caused by the overwhelming brightness of a star with respect to a dim terrestrial planet, astronomers have developed a novel technique known as *nulling interferometry*. This is performed by observing

Pioneer SETI

In 1959 Giuseppe Cocconi, former Director of CERN in Geneva, and Philip Morrison, of the Massachusetts Institute of Technology, proposed a search for extraterrestrial intelligence by listening for radio signals from outer space. The project planned to exploit the large radio telescope facilities around the world to analyse the incoming radio emissions at various frequencies. Peculiar behaviour – such as periodicities or other non-random events emerging from the background noise – would deserve attention as possible messages from an alien civilisation. But so far, no signs of extraterrestrial life have been detected.

Efforts have also been made to send our 'message in a bottle' across the Universe. In 1974 a coded radio signal was sent from Earth in the direction of the globular cluster M13 – to possibly be answered 50,000 years from now. The Pioneer and Voyager spacecraft have escaped the gravitational influence of the Solar System and are now travelling in interstellar space, taking with them gold plaques especially designed for communicating many aspects of our civilisation (Figure 11.8).

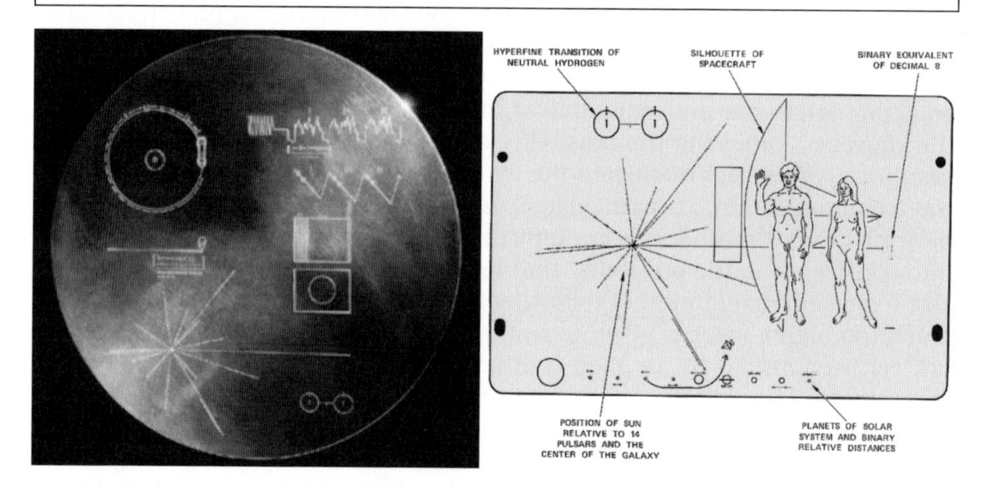

FIGURE 11.8. The gold plaque containing sounds and images from planet Earth, onboard the Voyager spacecraft (left), and the plaque carried by the Pioneer missions (right) showing the trajectory of the spacecraft beginning at the third planet from the Sun. (Courtesy NASA.)

the same target at the same time with telescopes in different locations, and by superposing the observations to cancel out the light of the star to reveal the faint objects orbiting it. This technique has already proved effective using ground-based telescopes, leading to the discovery of dust rings around some stars – a possible 'nursery' of planets.

Yet the real challenge is to perform nulling interferometry from space. The

European Space Agency is planning to place three space telescopes at the Lagrangian point L_2, which lies in the Sun–Earth direction behind our planet. Spaceflight dynamics is deeply involved, because the telescopes are to be carried on independent spacecraft that must be moving in formation, so that their relative distance is kept constant to a high degree of accuracy. This is the Darwin mission – so named because it will hopefully reveal the origin of life in the Galaxy, thus following in the footsteps of Charles Darwin (1809–1882), who provided the first convincing picture of the origin and evolution of life on our planet.

Appendix 1

Planetary data

	Semimajor axis (AU)	Eccentricity	Inclination (°)	Mass ($\times 10^{24}$ kg)	Equatorial radius (km)
Mercury	0.39	0.206	7.00	0.33	2,440
Venus	0.72	0.007	3.39	4.87	6,052
Earth	1.00	0.017	0.00	5.97	6,378
Mars	1.52	0.093	1.85	0.64	3,397
Jupiter	5.20	0.048	1.30	1,898.60	71,492
Saturn	9.54	0.054	2.49	568.50	60,268
Uranus	19.19	0.047	0.77	86.62	25,559
Neptune	30.07	0.009	1.77	102.78	24,764
Pluto*	39.48	0.249	17.14	0.1314	1,151

References
http://ssd.jpl.nasa.gov/txt/p_elem_t1.txt
http://ssd.jpl.nasa.gov/?planet_phys_par

*According to IAU Resolution 5A, 2006, Pluto is a 'dwarf planet'.

Appendix 2

Planetary satellite data

	Distance to planet (km)	Eccentricity	Inclination (degrees)	Mass (kg)	Mean radius (km)	Orbital resonances	Spin–orbit resonances
Earth							
Moon	384,400	0.0554	5.16	7.34767×10^{22}	1,737.5	Saros	1:1
Mars							
Phobos	9,380	0.0151	1.075	1.06975×10^{16}	11.1	1:4 Deimos	1:1
Deimos	23,460	0.0002	1.793	2.24351×10^{15}	6.2	4:1 Phobos	1:1
Jupiter							
Io	421,800	0.0041	0.036	8.93194×10^{22}	1,821.6	1:2 Europa; 1:4 Ganymede	1:1
Europa	671,100	0.0094	0.469	4.79984×10^{22}	1,560.8	1:2 Ganymede; 2:1 Io	1:1
Ganymede	1,070,400	0.0011	0.17	1.48186×10^{23}	2,631.2	2:1 Europa; 4:1 Io	1:1
Callisto	1,882,700	0.0074	0.187	1.07594×10^{23}	2,410.3		1:1
Amalthea	181,400	0.0031	0.388	2.06816×10^{18}	83.45		1:1
Thebe	221,900	0.0177	1.07	1.49867×10^{18}	49.3		
Adrastea	129,000	0.0018	0.054	7.49334×10^{15}	8.2		
Metis	128,000	0.0012	0.019	1.19893×10^{17}	21.5		

	Distance to planet (km)	Eccentricity	Inclination (degrees)	Mass (kg)	Mean radius (km)	Orbital resonances	Spin–orbit resonances
Himalia	11,461,000	0.1623	27.496	6.74401×10^{18}	85		
Elara	11,741,000	0.2174	26.627	8.69228×10^{17}	43		
Pasiphae	23,624,000	0.409	151.431	2.99734×10^{17}	30		
Sinope	23,939,000	0.2495	158.109	7.49334×10^{16}	19		
Lysithea	11,717,000	0.1124	28.302	6.29441×10^{16}	18		
Carme	23,404,000	0.2533	164.907	1.31883×10^{17}	23		
Ananke	21,276,000	0.2435	148.889	2.99734×10^{16}	14		
Leda	11,165,000	0.1636	27.457	1.09403×10^{16}	10		
Saturn							
Mimas	193,900	0.0193	1.572	3.79163×10^{19}	198.8	1:2 Tethys	1:1
Enceladus	238,040	0.0047	0.009	1.08054×10^{20}	252.3	1:2 Dione	1:1
Tethys	294,710	0.0001	1.091	6.17601×10^{20}	536.3		1:1
Dione	377,420	0.0022	0.028	1.09572×10^{21}	562.5		1:1
Rhea	527,070	0.001	0.331	2.309×10^{21}	764.5		1:1
Titan	1,221,870	0.0288	0.28	1.34553×10^{23}	2,575.5	3:4 Hyperion; 1:5 Iapetus	1:1
Hyperion	1,500,880	0.0274	0.63	5.54507×10^{18}	133	1:4 Iapetus; 4:3 Titan	Chaos
Iapetus	3,560,840	0.0283	7.49	1.8059×10^{21}	734.5	5:1 Titan	1:1
Phoebe	12,947,780	0.1635	175.986	8.28914×10^{18}	106.6		
Janus	151,500	0.0073	0.165	1.89731×10^{18}	90.4	1:1 Epimetheus	1:1
Epimetheus	151,400	0.0205	0.335	5.26033×10^{17}	58.3	1:1 Janus	1:1
Helene	377,420	0.0071	0.213	2.54774×10^{16}	16	1:1 Dione	
Telesto	294,710	0.0002	1.18	7.19361×10^{15}	12	1:1 Tethys	
Calypso	294,710	0.0005	1.499	3.5968×10^{15}	9.5	1:1 Tethys	
Atlas	137,700	0	0	2.09814×10^{15}	10		
Prometheus	139,400	0.0023	0	1.86734×10^{17}	46.8		
Pandora	141,700	0.0044	0	1.49118×10^{17}	40.6		
Pan	133,600	0	0	4.94561×10^{15}	12.8		

Uranus							
Ariel	190,900	0.0012	0.041	1.3533×10^{21}	578.9	3:5 Umbriel	1:1
Umbriel	266,000	0.0039	0.128	1.17196×10^{21}	584.7		1:1
Titania	436,300	0.0011	0.079	3.52637×10^{21}	788.9		1:1
Oberon	583,500	0.0014	0.068	3.01382×10^{21}	761.4		1:1
Miranda	129,900	0.0013	4.338	6.59414×10^{19}	235.8		1:1
Cordelia	49,800	0.0003	0.085	4.49601×10^{16}	20.1		
Ophelia	53,800	0.0099	0.104	5.39521×10^{16}	21.4		
Bianca	59,200	0.0009	0.193	9.29174×10^{16}	25.7		
Cressida	61,800	0.0004	0.006	3.43195×10^{17}	39.8		
Desdemona	62,700	0.0001	0.113	1.78342×10^{17}	32.0		
Juliet	64,400	0.0007	0.065	5.57505×10^{17}	46.8		
Portia	66,100	0.0001	0.059	1.68151×10^{18}	67.6		
Rosalind	69,900	0.0001	0.279	2.54774×10^{17}	36.0		
Belinda	75,300	0.0001	0.031	3.56683×10^{17}	40.3		
Puck	86,000	0.0001	0.319	2.89393×10^{18}	81.0		
Neptune							
Triton	354,800	0	156.834	2.13995×10^{22}	1,353.4		
Nereid	5,513,400	0.7512	7.232	3.08726×10^{19}	170.0		
Naiad	48,227	0.0004	4.746	1.94827×10^{17}	33		
Thalassa	50,075	0.0002	0.209	3.74667×10^{17}	41		
Despina	52,526	0.0002	0.064	2.09814×10^{18}	75		
Galatea	61,953	0	0.062	3.74667×10^{18}	88		
Larissa	73,548	0.0014	0.205	4.94561×10^{18}	97		
Proteus	117,647	0.0005	0.026	5.03553×10^{19}	210		
Pluto							
Charon	19,599	0.0022	96.151	1.61856×10^{21}	593		1:1

Reference
http://ssd.jpl.nasa.gov/?sat_elem, http://ssd.jpl.nasa.gov/?sat_phys_par

Appendix 3

Space missions

List of space missions operating at the distance of the Moon and beyond. For past or current missions the year denotes when operations began (such as the date of orbit insertion of Cassini around Saturn), and for future missions the launch date is given. If different missions with similar characteristics are named after the same programme (such as Surveyor), their total number is included in brackets. Only completely or partially successful missions have been included. The remarks include peculiar orbital configurations (such as horseshoe or halo orbits) and gravity assists (GA; V – Venus; E – Earth; M – Mars; J – Jupiter; and so on). EP indicates electric propulsion (rather than chemical propulsion). As a key, the names of the mission are formatted as follows:

Bold Past missions
Bold italics Current missions
Plain text Near-future approved missions
Italics Near-future approved missions under study

Target/ Year	Mission	Agency/ Country	Remarks
MOON			
1959–65	**Lunik 1–8**	USSR	Fly-by; orbiters
1961–65	**Ranger (3)**	NASA	Impact
1965–70	**Zond (8)**	USSR	Orbiters; Earth re-entry
1966–67	**Lunar Orbiter (5)**	NASA	Orbiters
1966–72	**Lunik 9–19**	USSR	Landers
1966–70	**Surveyor (7)**	NASA	Landers
1968–69	**Apollo 8, 10, 13**	NASA	Manned orbiter missions
1969	**Apollo 11, 12, 14**	NASA	Human exploration missions
1971–72	**Apollo 15, 16, 17**	NASA	Human exploration missions with rovers
1973	**Lunik 21/Lunokhod 2**	USSR	Lander; rover
1974	**Lunik 22**	USSR	Orbiter
1990	**Hiten**	ISAS	WSB transfer trajectory
1994	**Clementine**	NASA	Orbiter (extended mission to asteroid Geographos failed)

Target/ Year	Mission	Agency/ Country	Remarks
1998	**Lunar Prospector**	NASA	Orbiter
2004	**SMART-1**	ESA	Orbiter; electric peopulsion
2006	Lunar-A	JAXA	Orbiter; two penetrators
2007	Selene	JAXA	Main orbiter; two orbiting probes
2007	Chandrayaan-1	India	Orbiter
2008	Lunar Reconnaissance Orbiter	NASA	Orbiter
2009	*Moonrise*	NASA	Sample return
2010–20	*Robotic testbed missions (10)*	NASA	Landers; rovers; Earth re-entry
2015–20	*Human landings (5)*	International?	Manned exploration mission
(...)	*Lunar outpost*	International?	Permanent lunar outpost
MERCURY			
1974	**Mariner 10**	NASA	Three resonant fly-bys; VGA
2004	***Mercury Messenger***	NASA	Orbiter (2011); EVVGA
2012	*BepiColombo*	ESA/JAXA	Polar orbiter; magnetospheric orbiter (2014) EP
VENUS			
1961	**Venera 1**	USSR	Fly-by
1962	**Mariner 2**	NASA	Fly-by
1965–69	**Venera 3–6**	USSR	Atmospheric descent
1967	**Mariner 5**	NASA	Fly-by
1970/72	**Venera 7, 8**	USSR	Landers
1973	**Mariner 10**	NASA	Fly-by
1975–83	**Venera 9–14**	USSR	Orbiters; landers
1978	**Pioneer Venus**	NASA	Orbiter; four atmospheric probes
1983	**Venera 15, 16**	USSR	Orbiters for radar mapping of the surface
1985	**Vega 1, 2**	USSR	Release of atmospheric probes and balloons during fly-by
1990	**Galileo**	NASA	Gravity assist
1990	**Magellan**	NASA	Orbiter global radar mapping
2005	***Venus Express***	ESA	Orbiter
2008	Planet-C	JAXA	Orbiter
MARS			
1964–69	**Mariner 4, 6, 7**	NASA	Fly-by
1971	**Mars 2, 3**	USSR	Fly-by; impact
1971	**Mariner 9**	NASA	Orbiter
1974	**Mars 5, 6**	USSR	Orbiter; lander (failed)
1976	**Viking 1, 2**	NASA	Orbiters; landers
1989	**Phobos 2**	USSR	Orbiter; Phobos lander (failed)
1997	***Mars Global Surveyor***	NASA	Orbiter
1997	**Mars Pathfinder**	NASA	Microrover

2001	***Mars Odyssey***	NASA	Orbiter
2003	**Nozomi**	JAXA	Fly-by (orbital manoeuvre failed)
2003	***Mars Express***	ESA	Orbiter
2004	***Spirit***	NASA	Rover
2004	***Opportunity***	NASA	Rover
2005	***Mars Reconnaissance Orbiter***	NASA	Orbiter
2007	Phoenix	NASA	Rover
2009	Mars Science Laboratory	NASA	Rover
2009	Mars Telesat	NASA	Telecom orbiter
2011	ExoMars	ESA	Orbiter; lander; rover
2010–20	*Mars scouts (4)*	International?	Orbiters; rovers
2015	*Mars sample return*	International?	Sample return
2010–20	*Mars testbed missions (4)*	International?	Manned exploration preparatory missions
2033	*Mars human exploration*	International?	Human expedition
JUPITER			
1973	**Pioneer 10, 11**	NASA	Fly-by
1979	**Voyager 1, 2**	NASA	Fly-by
1995	**Galileo**	NASA	Orbiter; atmospheric probe; VEEGA
2008	JIMO	NASA	Icy Moons orbiter
2011	Juno	NASA	Polar orbiter
2020	Jupiter Orbiters	ESA	Micro-spacecraft missions
SATURN			
1979	**Pioneer 11**	NASA	Fly-by; JGA
1981	**Voyager 2**	NASA	Fly-by; JGA
2004	***Cassini–Huygens***	NASA/ESA	Orbiter; Titan probe; EVVJGA
URANUS			
1986	**Voyager 2**	NASA	Fly-by; JSGA
NEPTUNE			
1989	**Voyager 2**	NASA	Fly-by; JSUGA
PLUTO/TNO			
2006	***New Horizons***	NASA	Pluto (2015) and transneptunian objects fly-bys; JGA
SMALL BODIES			
1985	**ISEE3/ICE**	NASA	Comet Giacobini–Zinner fly-by
1986	**Vega 1, 2**	USSR	Comet Halley fly-by
1986	**Sakigake, Suisei**	ISAS	Comet Halley fly-by
1986	**Giotto**	ESA	Comet Halley fly-by

Target/ Year	Mission	Agency/ Country	Remarks
1991	**Gaileo**	NASA	Main belt asteroid Gaspra fly-by
1993	**Gaileo**	NASA	Main belt asteroid Ida fly-by
1997	**NEAR**	NASA	Main belt asteroid Mathilde fly-by
2000	**NEAR**	NASA	Near-Earth asteroid Eros orbiter. EGA
1997	**Deep Space 1**	NASA	NEA Braille (1998) and Comet Borrelly (2002) fly-bys; EP
2002	**Stardust**	NASA	Comet Wild 2 fly-by (2004); sample return (2006)
2003	***Hayabusa***	JAXA	NEA Itokawa rendezvous (2005); sample return (2006)
2004	***Rosetta***	ESA	Comet Churyumov–Gerasimenko orbiter; lander (2014); EMEEGA
2005	**Deep Impact**	NASA	Comet Tempel 1 fly-by; impact experiment
2006	Dawn	NASA	Main-belt asteroids (2011) Vesta and (2014) Ceres orbiter; EP
2009	*Don Quijote*	ESA	Orbiter; impact experiment; VGA
2015	Neosample return	ESA	Sample return from a NEO
ASTRONOMY/PHYSICS			
1978	**ISEE3/ICE**	NASA	Sun–Earth interaction; L_1 halo orbit
1990	**Ulysses**	ESA	Solar polar observatory; interplanetary space (Q>5 AU); JGA
1994	**Wind**	NASA	Interplanetary medium; L_1 halo orbit
1995	***Soho***	ESA	Solar observatory; L_1 halo orbit
1997	***ACE***	NASA	Interplanetary medium; L_1 halo orbit
1992	***Geotail***	ISAS/NASA	Geomagnetic tail; maximum apogee at 1.3 million km
2001	***Wmap***	NASA	Cosmology observatory; L_2 halo orbit
2003	***Spitzer***	NASA	Infrared observatory; horseshoe orbit
2006	Stereo	NASA	Two solar observatories; horseshoe orbit
2007	Herschel	ESA	Infrared observatory; L_2 halo orbit
2007	Planck	ESA	Cosmology observatory; L_2 halo orbit
2007	Kepler	NASA	Search for Earth-like planets; horseshoe orbit
2008	Lisa Pathfinder	ESA	Technology test for Lisa; L_1 halo orbit

2009	SIM	NASA	Space interferometry mission; horseshoe orbit
2011	Gaia	ESA	Astrometry; extrasolar planets; L_2 halo orbit
2011	JWST	NASA	James Webb Space Telescope; L_2 halo orbit
2012	Lisa	ESA/NASA	Fundamental physics; three s/c on horseshoe orbit
2012	DSCOVR	NASA	Deep Space Climate Observatory; L_1 halo orbit
2013	Solo	ESA	Solar observatory; interplanetary space ($q<0.3$ AU); EP
2014	*Darwin/TPF*	ESA/NASA	Search for habitable planets; six s/c constellation in L_2 halo orbit

Appendix 4

Internet resources

Astronomical data

Animations of the Solar System
http://cfa-www.harvard.edu/iau/Animations/Animations.html
Astdys (data on numbered and multi-opposition asteroids)
http://hamilton.dm.unipi.it/cgi-bin/astdys/astibo
Astronomical On-line Calculator (Java calculator with astronomical data)
http://www.astro.wisc.edu/~dolan/constants/calc.html
Comets (lists and plots)
http://cfa-www.harvard.edu/iau/lists/CometLists.html
Eclipses
http://sunearth.gsfc.nasa.gov/eclipse/eclipse.html
Minor Planet and Comet Ephemeris Service
http://cfa-www.harvard.edu/iau/MPEph/MPEph.html
Minor planets (lists and plots)
http://cfa-www.harvard.edu/iau/lists/MPLists.html
Natural satellites (mean orbital parameters)
http://ssd.jpl.nasa.gov/?sat_elem
Natural satellites (physical data)
http://ssd.jpl.nasa.gov/?sat_phys_par
Natural Satellites Service (information and tools on their dynamics)
http://lnfm1.sai.msu.su/neb/nss/index.htm
Nautical Almanac (including the history of HM Nautical Almanac Office)
http://www.nao.rl.ac.uk/
Planetary orbital elements
http://ssd.jpl.nasa.gov/txt/p_elem_tl.txt
Planetary physical properties
http://ssd.jpl.nasa.gov/?planet_phys_par
Planetary rings: Jupiter, Saturn, Uranus and Neptune
http://pds-rings.seti.org/
SIMBAD database (data on objects outside the Solar System)
http://cdsweb.u-strasbg.fr/Simbad.html
Small Bodies Node (NASA Planetary Data System)

http://pdssbn.astro.umd.edu/
Small-Body Orbital Elements (JPL DASTCOM database)
http://ssd.jpl.nasa.gov/sb_elem.html
Solar System exploration (a survey including data, images and videos)
http://sse.jpl.nasa.gov/index.cfm
Space calendar (space-related activities and anniversaries for the coming year)
http://www2.jpl.nasa.gov/calendar/

Astronomical images

European Space Agency multimedia gallery
http://www.esa.int/esaCP/index.html
Hubble Space Telescope observations (latest news releases)
http://hubblesite.org/newscenter/
Image Reduction and Analysis Facility (IRAF)
http://iraf.noao.edu/iraf-homepage.html
NASA NSSDC photograph gallery
http://nssdc.gsfc.nasa.gov/photo_gallery/
NASA Photojournal
http://photojournal.jpl.nasa.gov/index.html
NASA Picture of the Day
http://antwrp.gsfc.nasa.gov/apod/astropix.html
NASA Planetary Data System imaging node
http://pds-imaging.jpl.nasa.gov/

Books and journals

ADS (NASA Astrophysics Data System)
http://adswww.harvard.edu/
AMS Book Online (American Mathematical Society)
http://www.ams.org/online_bks/
Center for Retrospective Digitization (Göttingen University Library)
http://gdz.sub.uni-goettingen.de/en/index.html
Cornell Historic Math Book Collection
http://historical.library.cornell.edu/reroute_MATH.html
Electronic Newsletter for the History of Astronomy (ENHA)
http://www.astro.uni-bonn.de/~pbrosche/aa/enha/
Gallica: La Bibliothèque Numérique (Bibliothèque Nationale de France)
http://gallica.bnf.fr/

Dynamics and computer simulations

Jupiter! The Three Body Problem (animated simulation)
http://www.physics.cornell.edu/sethna/teaching/sss/jupiter/jupiter.htm
Restricted Three Body Problem (free software simulation (Jupiter))

http://www.physics.cornell.edu/sethna/teaching/sss/jupiter/Web/Rest3-Bdy.htm
Solar System Dynamics (information relating to all bodies orbiting the sun)
http://ssd.jpl.nasa.gov/
Three Bodies in Gravitation (simulation of a spacecraft within Earth–Moon gravity)
http://astro.u-strasbg.fr/~koppen/body/ThreeBody.html
Three Body Problem (equations of motion, fixed points and periodic orbits)
http://www.geom.uiuc.edu/~megraw/CR3BP_html/cr3bp.html

Space missions

ESA missions (in operation, under development, under study, and post-operation
http://sci.esa.int/science-e/www/area/index.cfm?fareaid=71
International Space Station (NASA human spaceflight)
http://spaceflight1.nasa.gov/station/
Lunar exploration (National Space Science Data Center)
http://nssdc.gsfc.nasa.gov/planetary/lunar/apollo_25th.html
NASA space science missions
http://science.hq.nasa.gov/missions/phase.html
Space Shuttle launches
http://science.ksc.nasa.gov/shuttle/missions/missions.html

Near-Earth objects

Kuiper Belt and Oort Cloud
http://seds.lpl.arizona.edu/nineplanets/nineplanets/kboc.html
Potentially hazardous asteroids (through to the end of the century)
http://cfa-www.harvard.edu/~graff/lists/Dangerous.html
Near-Earth Asteroids database (physical and dynamical properties)
http://earn.dlr.de/nea/
NEO Page
http://cfa-www.harvard.edu/iau/NEO/TheNEOPage.html

Extrasolar planetary systems

California and Carnegie Planet Search
http://exoplanets.org/
Extrasolar Planet Encyclopaedia
http://vo.obspm.fr/exoplanetes/encyclo/
Geneva Extrasolar Planet Search
http://obswww.unige.ch/~udry/planet/planet.html

Societies and other organisations

American Astronomical Society
http://www.aas.org/
European Space Agency (ESA)
http://www.esa.int/esaCP/index.html
ESA Science (Science and technology programme)
http://sci.esa.int/science-e/www/area/index.cfm?fareaid=1
International Astronomical Union (IAU)
http://www.iau.org/
IAU Commission 7 on Celestial Mechanics and Dynamical Astronomy
http://copernico.dm.unipi.it/comm7/
IAU Minor Planet Center
http://cfa-www.harvard.edu/iau/mpc.html
Italian Society of Celestial Mechanics and Astrodynamics (SIMCA)
http://www.mat.uniroma2.it/simca/
Jet Propulsion Laboratory (JPL)
http://www.jpl.nasa.gov/
National Areonautics and Space Administration (NASA)
http://www.nasa.gov/
SETI Institute (Search for Extraterrestrial Intelligence)
http://www.seti.org/
Spaceguard Foundation
http://spaceguard.esa.int/

Glossary

Accretion disk A disk-shaped region composed of dust and gas feeding the gravitational accretion of a celestial body.

Apocentre The point of maximum distance from the focus of a celestial body travelling on an elliptical orbit.

Argument of pericentre The angular distance (denoted by the Greek letter ω) of the pericentre from the line of nodes, defining the orientation of an orbit in its plane. If the orbit is circular, the argument of pericentre is undetermined.

Asteroid From the Greek αστεροειδης ('star-like'). An irregularly-shaped rocky body orbiting the Sun. Three asteroid populations are known: main-belt, near-Earth and Trojan, characterised by different orbital regimes.

Asteroid belt The region of the Solar System between Mars and Jupiter, where main-belt asteroids reside.

Astronomical Unit (AU) The standard unit of measurement denoting the distances of celestial objects in the Solar System. 1 AU is the Earth–Sun mean distance: 149,597,870 km.

Big Bang A cosmological theory according to which the Universe formed between 12 and 14 billion years ago, from a giant explosion concentrated in a singular point with infinite density.

Chaos From the Greek χαος, denoting primordial emptiness. The term is used to indicate the extreme sensitivity of a trajectory to the initial conditions, which implies unpredictable dynamical behaviour in the corresponding dynamical system.

Closure error A procedure for evaluating the internal accuracy of a numerical method which consists of moving back and forth along the same branch of a trajectory, with the aim of checking how closely the initial conditions are restored.

Coma The bright region surrounding the nucleus of an active comet.

Comets Irregularly-shaped icy bodies on widely different heliocentric orbits. When a comet approaches the Sun within less than 2 AU the ices sublimate, and the dust and gas ejected into space form the coma and tail.

Conjunction An inferior planet (with a semimajor axis smaller than Earth's) is in conjunction when it is aligned with the Earth and the Sun. A superior planet (with a semimajor axis greater than Earth's) is in conjunction when it is aligned with the Earth and the Sun on the opposite side of the Sun.

Constellations Groups of physically or perspectively nearby stars identified by a name associated with a mythological character.

Direct motion The motion along an orbit in counterclockwise direction, as followed by the vast majority of Solar System objects. The opposite of retrograde motion.

Edgeworth–Kuiper Belt A disk-shaped region extending beyond the orbit of Neptune, where a whole population of primordial celestial bodies resides.

Eccentricity The orbital parameter measuring the deviation of an ellipse from a circle. It varies between 0 (a circle) and 1 (a parabola).

Ecliptic plane The plane in which the trajectory of the Earth around the Sun lies. All planets move on trajectories with orbital planes remarkably close to the ecliptic.

Ellipse A planar curve characterised by the property that the sum of the distances of a generic point on the curve from two fixed points (foci) is constant. When the foci coincide, the ellipse degenerates into a circle. The orbital path of most celestial bodies closely resembles an ellipse.

Equilibrium position The point at which the sum of the forces acting within the system is zero. An equilibrium position can be either stable or unstable.

Equinoxes From the Latin *aequa nox* ('equal night'), marking the time of the year (around 23 September and 21 March) when the orientation of the spin axis of the Earth, with respect to Earth's position along the orbit, results in days and nights of equal length.

Escape velocity The lowest value of the initial velocity of an object departing from the surface of a celestial body required to escape the body's gravitational attraction. Earth-escape velocity is 11.2 km/s, while for the Moon it is 2.4 km/s.

Extrasolar planet (exoplanet) A planet orbiting a star other than the Sun.

Exobiology The study of the conditions for the development of life in the Universe.

Fly-by A mission profile by which a spacecraft encounters a celestial body at the intersection of the spacecraft trajectory with that of the target. Fly-bys take place at high relative velocity and allow short-duration observations.

Gravitational constant The fundamental constant in Newton's law of gravitation.

Gravity assist A close encounter of a spacecraft with a massive celestial body in order to exploit the gravity of the body in redirecting the spacecraft's motion.

Halo orbit A periodic and quasi-periodic motion around the collinear Lagrangian points. The trajectories appear as haloes surrounding the celestial bodies involved.

Hyperbola A planar curve characterised by the property that the difference of the distances of a generic point on the curve from two fixed points (foci) is constant.

Inclination Angular separation between the orbital plane of a celestial body and a given reference plane (such as the ecliptic).

Integrable system A system in which dynamics is provided by periodic or quasi-periodic trajectories. The equations of motion of an integrable system can be solved exactly.

KAM theory Given an integrable system, KAM theory provides for the persistence of quasi-periodic motions under a small perturbation. It can be applied, with general assumptions on the integrable system and on the frequency of motion, and yields a constructive algorithm to evaluate the strength of the perturbation, ensuring the existence of quasi-periodic trajectories.

Kepler's laws Originally stated for describing the motions of the planets. They rule the motion of any pair of isolated celestial objects, and represent a first-order approximation to the motion of most Solar System objects.

Lagrangian points Stationary solutions of the three-body problem. Three of them are referred to as 'collinear', because they all lie on the line joining the other two bodies; while the remaining two are 'triangular', because they form equilateral triangles with their position.

Libration Oscillations of an orbital parameter around an equilibrium value. Practical examples are the wobbling of the Earth's spin axis around the mean value of the obliquity, and the oscillation of the semimajor axis of an orbit around a resonant value.

Line of nodes The line traced by the intersection between two orbital planes. When assuming a reference plane (such as the ecliptic) it defines the orientation of the plane of an orbit in space. If the inclination is zero, the line of nodes is indeterminable.

Light-year The distance travelled by electromagnetic radiation (light) in empty space during one year: 9.46×10^{12} km (about 10,000 billion km). It is used to measure distances on a galactic scale.

Longitude of the ascending node The angular distance (denoted by the Greek letter Ω) of the ascending node from a reference direction.

Long-period comets Comets having a period of revolution longer than 200 years. Their origin can be traced to the Oort Cloud.

Magnitude A measure of the brightness of an object on a logarithmic scale. Each unit corresponds to an increase or a decrease in brightness by a factor of 2.512. *Apparent magnitude* denotes the brightness of a celestial body as seen from Earth, while *absolute magnitude* is the calculated brightness at a standard distance. Naked-eye observations reach to 6th magnitude or a little fainter, while state-of-the-art telescopes are able to detect objects fainter than magnitude 26.

Manoeuvre An artificially generated perturbation on the motion of a spacecraft produced by firing its onboard propulsion system.

Minimum Orbit Intersect Distance (MOID) The closest approach distance between two trajectories in three-dimensional space.

N-body problem A gravitationally interacting dynamical system composed of an arbitrary number of bodies.

Near-Earth asteroids (NEA) Asteroids with perihelia close to or less than Earth's aphelion. They are fragments of catastrophic collisions in the asteroid main belt, arriving in the inner Solar System on chaotic orbits.

Near-Earth objects (NEO) Natural or artificial bodies (NEAs, most comets, and space debris) with trajectories approaching that of Earth.

Nearly-integrable system A problem that cannot be solved exactly, but which can be modelled as a small perturbation of an integrable system. For example, the motion of an asteroid can be considered as an elliptic trajectory (the integrable system), slightly perturbed by the gravitational attraction of Jupiter.

Netwon's law (of gravitation) This states that two masses attract each other with a force which increases with the values of the masses, and decreases with the distance between them. Kepler's laws can be straightforwardly derived from this basic principle.

Node (nodal distance) The distance of the point of interception of an elliptical trajectory with a given plane. Each orbit has two nodes, ascending or descending, depending on whether the plane is crossed from south to north or *vice versa*. When an asteroidal orbit and the ecliptic are involved, the node marks the minimum achievable distance between the Earth and the asteroid.

Nominal trajectory The predicted trajectory of a spacecraft. Mid-course and station-keeping manoeuvres are performed in order to keep the true trajectory of a spacecraft as close as possible to the nominal trajectory.

Numerical methods Computer-aided procedures for studying the evolution of a dynamical system.

Obliquity The angle between the rotation axis and the perpendicular to the orbital plane. (Alternatively, the orbital plane is replaced with the ecliptic plane.)

Occultation A celestial body passing in front of another in the observer's line of sight (for example, the Moon passing in front of Jupiter). Stellar occultations are used for measuring the size of small celestial bodies and for investigating the structure of planetary rings.

Oort Cloud The spherical region 50,000–100,000 AU from the Sun, considered as the source region of long-period comets.

Opposition A superior planet (with a semimajor axis greater than the Earth's) is in opposition when it is directly on the opposite side of the Earth with respect to the Sun.

Parabola A planar curve with particular points at the same distance from a fixed point (the focus) and a fixed line (the directrix).

Parsec A measure of distance, equal to about 3.3 light-years. It is defined as the distance at which two objects 1 AU apart are separated by an angle of 1 arcsecond.

Payload Spacecraft onboard subsystems dedicated to the achievement of the scientific or technological purposes of a mission (such as the high-resolution camera onboard an exploratory mission and the radio signal amplifiers of a telecommunication satellite).

Pericentre The minimum distance from the focus of a celestial body on an elliptical orbit.

Periodic orbit A trajectory repeating itself after a given interval of time (the period).

Perturbation theory A mathematical theory for providing an approximate solution to the equations of motion of a nearly-integrable system.

Potentially hazardous asteroids (PHA) Asteroids that will approach the Earth to within 0.05 AU (7.5 million km – about twenty times the distance of the Moon) in the near future (100–200 years).

Planetesimal A kilometre-size body resulting from the early phases of planetary accretion. Many asteroids and comets are thought to be planetesimals formed in different regions of the Solar System.

Primitive bodies Planetesimals which have undergone minor physical changes since their formation, thus retaining pristine material from the early Solar System. Most primitive bodies can be found among the meteorites, asteroids, comets and TNOs.

Precession The motion of certain characteristic parameters of a dynamical system, such as the conic 'wobble' described by the spin–axis around a reference axis, or the slow motion of the pericentre on the orbital plane.

Quasi-integrable system The same as nearly-integrable system.

Quasi-periodic motion A solution of the equations of motion that approaches indefinitely close to its initial conditions at regular intervals of time, though never exactly retracing itself.

Rendezvous A mission profile in which a celestial body is approached with a relative velocity low enough to allow gravitational capture by the target. In the astronautical sciences the term denotes the approach of two spacecraft for docking.

Resonance A commensurability among the periods of motion of two or more celestial bodies. The most common resonances are the mean motion resonances, which involve the revolution periods of different celestial bodies, and the spin–orbit resonances between the revolution and the rotation periods of the same celestial body.

Restricted three-body problem A special case of the three-body problem, where it is assumed that one of the bodies has a mass small enough that it does not perturb the motion of the other two bodies. Typical restricted problems are the Sun–Jupiter–asteroid and the Sun–planet–spacecraft systems.

Retrograde motion Motion along an orbit in clockwise direction (the opposite of direct motion).

Semimajor axis The distance between the centre of the ellipse and its apocentre (or pericentre). When orbital motion is involved it represents the mean distance from the focus.

Short-period comets Comets having a revolution period less than 200 years. They can be dynamically separated into Jupiter-family comets (JFC), which move on moderately eccentric and inclined orbits, and Halley-type comets, which possess elongated high-inclination or retrograde orbits.

Singularity An infinity occurring in a mathematical function or in a physical situation. For example, a division of a number by zero is a mathematical singularity. A collision between two celestial bodies is a physical singularity.

Solstice From the Latin *solstitium* ('Sun standing still'). The time of the year when the orientation of the spin axis of the Earth, with respect to its position along the orbit, allows a minimum (winter solstice) or a maximum (summer solstice) amount of daylight.

Step-by-step A numerical procedure used for computing a trajectory. The evaluation of the dynamical evolution of a system in subsequent time-steps.

Synchronous resonance A peculiar spin–orbit resonance occurring whenever the ratio between the revolution and rotation periods of a celestial body is unity. When applied to natural satellites it implies that the satellite always points the same hemisphere toward the host planet.

Three-body problem A system composed of three celestial bodies (such as Sun–planet–satellite). In its general form, no analytical solution to the equations of motion can be found.

Transneptunian objects (TNO) Bodies gravitationally bound to the Sun and moving beyond the orbit of Neptune. Edgeworth–Kuiper Belt objects and Oort Cloud comets are examples of TNOs.

Two-body problem A system composed of two celestial bodies (such as Sun–planet or planet–satellite) bound by gravitational attraction. The equations of motion of the two-body problem are integrable, and the solutions are described by Kepler's laws.

Yarkowsky effect A small perturbation of the motion of a celestial object due to the non-uniform heating of its surface as a consequence of its rotation.

Bibliography

Adams, D., *The Hitchhiker's Guide to the Galaxy*, Wings Books, 1996.

Airy, G.B., *An Elementary Explanation of the Principal Perturbations in the Solar System*, NEO Press, Ann Arbor, 1969.

Bell, E.T., *Men of Mathematics*, Simon and Schuster, New York, 1950.

Bertotti, B. and Farinella, P., *Physics of the Earth and the Solar System*, Kluwer Academic Publishers, Dordrecht, 1990.

Brouwer, D. and Clemence, G.M., *Methods of Celestial Mechanics*, Academy Press London, 1961.

Caro, T.L., *Della Natura*, Sansoni, Firenze, 1969.

Celletti, A. and Chierchia, L., 'KAM Stability and Celestial Mechanics', to appear in *Memoirs AMS.*

Celletti, A., Ferraz-Mello, S. and Henrard, J. (eds.), Modern Celestial Mechanics: From Theory to Applications, *Celestial Mechanics and Dynamical Astronomy*, Special Issue, 83, 2002.

Celletti, A., Milani, A. and Perozzi, E. (eds.), Second Italian Meeting on Celestial Mechanics, *Planetary and Space Science*, Special Issue, 11/12, 1998.

Coletta, A. and Gandolfi, G., *Il Secondo Big Bang: Esplosioni Cosmiche ai Confini dell'Universo*, CUEN, Napoli, 2000.

Diacu, F. and Holmes, P., *Celestial Encounters: The Origins of Chaos and Stability*, Princeton University Press, 1996.

Dreyer, J.L.E., *A History of Astronomy from Thales to Kepler*, Dover Publication, 1953.

Elliott, J. and Kerr, R., *Rings*, MIT Press, 1987.

EXPLOSPACE Workshop: Space Exploration and Resources Exploitation. ESA WPP-151, 1999.

Flammarion, C., *Les Terres Du Ciel*, C. Marpon et E. Flammarion editeurs, Paris, 1884.

Fresa, A., *La Luna: Movimenti, Configurazioni, Influenze e Culto*, Hoepli, Milano, 1952.

Grossman, N., *The Sheer Joy of Celestial Mechanics*, Birkhauser, 1996.

Gurzadyan, G., *Theory of Interplanetary Flights*, Gordon and Breach, Amsterdam, 1996.

Invernizzi, L., Manara, A. and Sicoli, P., *L'Astronomo Valtellinese Giuseppe Piazzi e la Scoperta di Cerere*, Fondazione Credito Valtellinese, Sondrio, 2001.

Life in the Universe, *Scientific American*, Special Issue, Freeman and Company, New York, 1995.

Maio, R. de., *Pulcinella e la Luna*, Casa Editrice Fausto Fiorentino, Napoli, 1992.

Manara A., *La Terra nel Mirino: Asteroidi e Probabilità di Collisione*, Il Castello, Trezzano sul Naviglio, 2003.

Marchal, C., *The Three-Body Problem*, Elsevier, Amsterdam, 1990.

Montenbruck, O. and Pfleger, T., *Astronomy on the Personal Computer*, Springer, 2004.

Morbidelli, A., *Modern Celestial Mechanics: Dynamics in the Solar System*, CRC Press, 2002.

Morrison, D., *Voyages to Saturn*, NASA SP-451, 1982.

Murray, C.D. and Dermott, S.F., *Solar System Dynamics*, Cambridge University Press, 1999.

Neugebauer, O., *The Exact Sciences in Antiquity*, Dover Publications, New York, 1969.

Öpik, E., *Interplanetary Encounters: Close-Range Gravitational Interactions*, Elsevier, Amsterdam, 1976.

Orbital Debris: A Technical Assessment, National Academy Press, Washington DC, 1995.

Pannekoek, A., *A History of Astronomy*, Dover Publications, New York, 1989.

Peterson, I., *Le Chaos dans le Systeme Solaire*, Pour la Science, Paris, 1995.

Poincaré, H., *Les Methodes Nouvelles de la Mechanique Celeste*, Gauthier-Villars, Paris, 1892.

Poincaré, H., *Science and Hypothesis*, Dover Publications, New York, 1952.

Prussing, J. and Conway, B., *Orbital Mechanics*, Oxford University Press, 1993.

Roy, A.E., *Orbital Motion*, Adam Hilger, 1988.

Sobel, D., *The Planets,* Viking, New York, 2005.

Stern, A., *Our Worlds: The Magnetism and Thrill of Planetary Exploration*, Cambridge University Press, 1999.

Szebehely, V., *Adventures in Celestial Mechanics: A First Course in the Theory of Orbits*, University of Texas Press, 1989.

Index

Printing: Mercedes-Druck, Berlin
Binding: Stein + Lehmann, Berlin